AF614281

G. Gautschi

Piezoelectric Sensorics

Springer-Verlag Berlin Heidelberg GmbH

G. Gautschi

Piezoelectric Sensorics

Force
Strain
Pressure
Acceleration and Acoustic Emission Sensors
Materials and Amplifiers

With 175 Figures

Springer

Dipl.-Ing. ETH Gustav Gautschi
Oetlisbergstr. 25
CH-8053 Zürich
Switzerland
e-mail: gustav.gautschi@alumni.ethz.ch

DOI 10.1007/978-3-662-04732-3

Library of Congress Cataloging-in-Publication-Data

Gautschi, G. (Gustav), 1936-
Piezoelectric sensorics : force, strain, pressure, acceleration and acoustic emission sensors, materials and amplifiers / G. Gautschi.
p. cm.
Rev. and enlarged translation of chapters 6 - 12 of German text by Jan Tichý and Gustav Gautschi.
Includes bibliographical references and index.

1. Piezoelectric transducers. I. Title.
TK7872.P54 G38 2002
681'.2–dc21

http://www.springer.de

Originally published by Springer-Verlag Berlin Heidelberg New York in 2002.
MyCopy version of the original edition 2002

Typesetting and Cover Design: medio Technologies AG, Berlin
Printed on acid free paper SPIN: 10793045 62/3020/M – 5 4 3 2 1 0
www.springer.com/mycopy

Preface

Sensors are the key to life and survival – and to the success of modern technology. Nature has provided living creatures with a wealth of sensors for a variety of measurands, such as light, sound, temperature, speed, motion, distance, force, pressure, acceleration, odor and so on – sensors, whose performance and specifications have often not been matched yet by man-made devices. Even at today's high level of electronics and information technology, sensors remain the crucial and decisive interface needed to reliably relate phenomena occurring in the environment to corresponding electric signals that can be processed to obtain the desired information and subsequent correct reaction of systems.

Although the literature on sensors is extremely vast, there is one type of sensors which so far has received little attention: the piezoelectric sensor. Certainly, most handbooks on measurement mention briefly this type of sensor yet there is not a single book in the English language dedicated entirely to piezoelectric sensors and giving a reasonably complete overview. There are only the books by [Gohlke 1955 and 1959] and [Tichý and Gautschi 1980], all in German.

The information and explanation given on piezoelectric sensors in measurement handbooks is – regrettably – rather terse and often quite limited. In particular there are certain prejudices and even misconceptions about piezoelectric sensors that have been perpetuated over many years and still persist today, such as: "piezoelectric sensors can only be used for dynamic measurements and are usually limited to measuring vibration". My practical experience of over 30 years in the field and close contact with countless users of sensors worldwide, has confirmed again and again that there is indeed a widespread lack of full understanding of the nature of piezoelectric sensors and their applications, both in academia and industry.

Therefore the goal of this book is to give a comprehensive overview of piezoelectric sensors, their characteristics and their proven practical applications. The book is not intended to be an instruction on how to build piezoelectric sensors. Rather, their properties and their "behavior" as sensing devices are presented with the aim to enable the reader to make a better and more objective choice when looking for a suitable sensor to fulfill a given measuring task. For each type of sensor, some exemplary applications are described to illustrate the possibilities offered and to give the reader ideas for his own work.

In chapter 1, a brief comparison of the piezoelectric sensing principle with other types of sensors illustrates the possibilities offered. The historic background highlighted in chapter 2 describes a number of interesting facts which are not so well known generally. The piezoelectric materials and their physical properties currently used in sensors are presented in chapter 3. An important but often neglected subject is the terminology for correctly defining and comparing the characteristics (specifications) of sensors, the topic of chapter 4. Chapter 5 gives an overview of the parts and the basic design of sensors. Sensors for the measurands *force, strain, pressure, acceleration* and *acoustic emission* are presented in chapters 6 to 10. These chapters are intentionally well illustrated to show the reader the great variety of applications existing for piezoelectric sensors.

Chapter 11 treats the amplifiers used with piezoelectric sensors and their particular nature. Here again there are still widespread misconceptions about these amplifiers because the output signal of piezoelectric sensors – electric charge – is not so often encountered in other fields of measuring electronics.

The number of references cited had to be kept within reasonable limits and is far from exhaustive. Readers may also refer to the included list of manufacturers of piezoelectric sensors from whom more information can be obtained.

The first book written together with Jan Tichý [Tichý and Gautschi 1980] covered both, the theory of the physics of crystals, and the piezoelectric sensors and their applications. Practical considerations prompted us to split the book into two volumes. This book corresponds to the second part (chapters 6 ... 12) of the previous book, considerably enlarged and brought up to date. The chapters on the measurands *strain* and *acoustic emission* are new. The first part of the previous book authored by Jan Tichý (chapters 1 ... 6) has also been rewritten, enlarged and updated by Jirí Erhart, Jan Fousek, Václav Janovec, Jana Prívratská and Jan Tichý [Erhard et al 2002]. Therefore, the two new volumes are complementary, but can be used separately, too.

I am highly indebted to a number of friends and colleagues who significantly contributed to the manuscript. Claudio Cavalloni reviewed the manuscript and wrote the chapter 3.5 on the new materials of the CGG group while Roland Sommer contributed to the chapter 3.7 on piezoelectric ceramics and thin films. Wolfgang Wallnöfer and Peter Worsch provided the chapter 3.4 on gallium orthophosphate. The manuscript was also critically reviewed and commented on by Christoph Gossweiler, Karl-Heinz Martini and Jan Tichý, co-author of our first book in German.

My thanks go to a number of sensor manufacturers who generously contributed material that serves to exemplify the vast range and variety of applications. Throughout the project, the staff of the Springer-Verlag offered an excellent cooperation and continuous support for which I am especially grateful.

Although this book is based on the previous German edition and has been enlarged, it is a first attempt to cover the entire field of piezoelectric sensors in a single volume. There is certainly room for improvement and I greatly welcome comments, criticism and suggestions (gustav.gautschi@alumni.ethz.ch) to consider for further enhancing a next edition.

Zurich, November 2001

Gustav H. Gautschi

Contents

Symbols, Quantities and Units

Note 1: Quantities, units and symbols are represented in this book in accordance with the Standards ISO 31-0 ... ISO 31-13:1992 and ISO 1000:1992 [ISO 1993], valid at the time of printing this book.
In particular, the reader should note that ISO 31-0:1992, Section 3.3.2, states:

> **The decimal sign is a comma on the line.**
> *If the magnitude of the number is less than unity, the decimal sign should be preceded by a zero. Note: In documents in the English language, a dot is often used instead of a comma. If a dot is used, it should be on the line. In accordance with an ISO Council decision,* ***the decimal sign is a comma in ISO documents.***

Therefore, it was decided to use a comma as decimal sign in this book, too. In accordance with section 3.3.1 of ISO 31-0:1992, the reading of numbers with many digits is facilitated by separating them in groups of three, *separated by a small space, and never by a comma or a point, nor by any other means.*

Note 2: For indicating a range of values, the mathematical sign "..." is used, i.e. *a* ... *b* means *from a to b, limits included.*

Symbol	Quantity	Unit
A	factor of proportionality	
	amplitude	m
	area, surface	m^2
a	thickness	m
	acceleration	ms^{-2}
a, b, c	crystallographic axis	
b	width	m
C	spring constant	Nm^{-1}
	capacitance	F
c_{ijkl}, $c_{\lambda\mu}$	modulus of elasticity	Nm^{-2}
D	electric flux density	Cm^{-2}
d_h	coefficient of the hydrostatic piezoelectric effect	CN^{-1}
$d_{i\lambda}$	piezoelectric coefficient	CN^{-1}
E	Young's modulus of elasticity	Nm^{-2}

E	electric field strength	Vm^{-1}
E_l	electric field strength of the local field	Vm^{-1}
e	base of the natural logarithms (e=2,718 281 ...)	
e	elementary charge ($e = (1{,}602\,177\,33 \pm 0{,}000\,000\,49) \cdot 10^{-19}$ C)	
$e_{i\lambda}$	piezoelectric modulus	Cm^{-2}
F	force	N
f	frequency	Hz
G	shear modulus	Nm^{-2}
g	acceleration of free fall	ms^{-2}
g_n	standard acceleration of free fall {$g_n = 9{,}806\,65$ m/s^2 (exactly), usually simply designated as g}	
$g_{i\lambda}$	piezoelectric coefficient	m^2C^{-1}
$g_{i\lambda\mu}$	electroelastic coefficient	Cm^2N^{-2}
$\boldsymbol{H}$	magnetic field strength	Am^{-1}
$h_{i\lambda}$	piezoelectric modulus	NC^{-1}
I	electric current	A
k	spring constant	Nm^{-1}
l	length	m
M	moment	Nm
m	mass	kg
n_1, n_2, n_3	integers	
O	origin of coordinates	
P	point	
$\boldsymbol{P}$	electric polarization	Cm^{-2}
P_s	spontaneous polarization	Cm^{-2}
p	pressure	Pa
p_i	pyroelectric coefficient	$Cm^{-2}K^{-1}$
Q	electric charge	C
q_i	pyroelectric modulus	m^2C^{-1}
$q_{ij\lambda}$	electrostrictive coefficient	m^2V^{-2}
r	radius	m
$r_{\lambda k}$	electrooptic constant	mV^{-1}
s	displacement	m
$s_{ijkl}, s_{\lambda\mu}$	elasticity coefficients	m^2N^{-1}
T	torque	Nm
U	tension	V
u	displacement	m
V	volume	m^3
v	speed	ms^{-1}
	gain (amplification factor)	1
W	work	J
w	energy density	Jm^{-3}
X_1, X_2, X_3	coordinate axes, material or Lagrange's coordinates	
x_1, x_2, x_3	coordinate axes, spatial or Euler's coordinates	
x, y, z	cartesian coordinate axes	
α	damping constant	kgs^{-1}
$\alpha_{ij}, \alpha_\lambda$	thermal expansion coefficients	K^{-1}

Δ	logarithmic decrement	
δ	damping coefficient	s^{-1}
ε	permittivity	Fm^{-1}
ε_0	electric constant, permittivity of vacuum (e_0 = 8,854 188 pF/m)	
ε_{ijk}	electrooptic coefficient	CV^{-2}
Θ	angle	
Θ	temperature	K, °C
Θ_C	Curie temperature	K, °C
δ	angle	
	damping coefficient	
ν	Poisson ratio	1
ξ, η, ζ	angles of rotation about the x_1, x_2, x_3 axes	
Π	hydrostatic pressure	N/m^2
$\Pi_{\lambda\mu}$	piezooptic constant	m^2N^{-1}
π	Ludolf number (π = 3,141 592 6...)	
π	pyroelectric coefficient	$Vm^{-1}K^{-1}$
π	piezomagnetic coefficient	$A^{-1}m$
ρ	density (mass)	kgm^{-3}
	density (charge)	Cm^{-3}
ρ_i	pyroelectric modulus	$V^{-1}m$
σ	normal stress	Nm^{-2}
σ_{ij}, σ_l	thermal expansion modulus	Nm^{-2}
τ	shear stress	Nm^{-2}
τ	time constant	s
τ_{ij}, τ_λ	thermal tension coefficient	$Nm^{-2}K^{-1}$
φ	phase angle	
ω	angular frequency	s^{-1}

Decimal prefixes (SI)

Factor	Name	Symbol	Factor	Name	Symbol
10^{24}	yotta	Y	10^{-3}	milli	m
10^{21}	zetta	Z	10^{-6}	micro	µ
10^{18}	exa	E	10^{-9}	nano	n
10^{15}	peta	P	10^{-12}	pico	p
10^{12}	tera	T	10^{-15}	femto	f
10^{9}	giga	G	10^{-18}	atto	a
10^{6}	mega	M	10^{-21}	zepto	z
10^{3}	kilo	k	10^{-24}	yocto	y
10^{2}	hecto	h			
10	deca	da			
10^{-1}	deci	d			
10^{-2}	centi	c			

1 Introduction

Measure what is measurable, and make measurable what is not so.

Galileo Galilei, 1564 – 1642

In physical science the first essential step in the direction of learning any subject is to find principles of numerical reckoning and practicable methods for measuring some quality connected with it.

I often say that when you can measure what you are speaking about, and express it in numbers, you know something about it; but when you cannot measure it, when you cannot express it in numbers, your knowledge is of a meagre and unsatisfactory kind; it may be the beginning of knowledge, but you have scarcely in your thoughts advanced to the state of science, whatever the matter may be.

William Thomson (Lord Kelvin), 1824 – 1907

Door meten tot weten. (To knowledge by measurement.)

Onnes Kammerlingh, 1853 – 1926
Dutch physicist

A measurement is the experimental determination of the magnitude of a physical quantity by comparing it with the corresponding unit of measurement. However, most measurements are made indirectly by exploiting physical effects in suitable measuring devices such as transducers or sensors. Most often the quantity to be measured is converted by a sensor into an output quantity of different nature that then can be amplified and recorded as required. A particularly convenient type of output signal is an electric quantity, and one speaks of an electric measuring method.

In electric measuring of mechanical quantities, a mechanical input quantity generates or controls an electric signal that represents, as output, the measured quantity. The assignment of the numerical indication to the measured quantity is done in two steps: first, the mechanical quantity is converted into an electric signal, and then the electric signal is amplified, measured and recorded.

Although in the 19th century already, very short or fast phenomenon were successfully measured in the laboratory by means of electric methods, the electric measuring of mechanical quantities started to gain importance only towards the middle of the 20th century. A decisive factor was the rapid development of the electric amplifier technique, first with vacuum tubes and then with semiconductors.

Today, at the onset of the 21st century, electric measuring of mechanical quantities has become an indispensable tool not only in research but in industry (process monitoring and control, etc.), too. Indeed, the possibilities offered by modern microprocessors and electronic data processing can be exploited fully only with suitable sensors for capturing the mechanical quantities.

Mechanical quantities that can be measured electrically are primarily displacement, velocity, force, torque, strain, pressure, sound, acceleration and acoustic emission. A number of physical effects can serve for converting a mechanical quantity into an electric output signal. The most important ones are a change in resistance (strain gage, piezoresisitive and potentiometric sensors), a change in capacitance or in inductance, and a change in polarization (piezoelectric effect). Other effects used for transduction are e.g. optical and chemical effects.

Basically there are two types of sensors: *active* and *passive* sensors. A sensor is called *active* if no external source of power is required for measuring. Piezoelectric sensors are of the active type because the electric charge yielded by their transduction element (a piezoelectric material) in response to a mechanical load can be indicated e.g. with a gold foil electrometer which does not need an external power source (see 11.2.1). Most other sensors are of the *passive* type, i.e. they do not directly yield an output, rather they – passively – change their electric properties (change in resistance, capacitance or inductance) as a function of the measurand. Such a change can only be detected by applying an external source of power which will reveal the output signal in the form of a change in electric current or voltage.

Overviews over the various sensor systems in general use can be found e.g. in [Bush-Vishniac 1999; Bao 2000; Doebelin 1989; Frank 1996; Fraden 1997; Neubert 1975; Norton 1969 and 1982; Noltingk 1996; Pflier 1978; Profos 1994; Sydenham 1989; Sydenham et al 1992; Webster 1999]. This book concentrates on piezoelectric sensors only and therefore we give here only a brief comparison of the main principles used. Piezoelectric sensors for force, torque, strain, pressure, acceleration and acoustic emission are covered in this book. A comparison is made with resistive, capacitive and inductive sensors which are also used to measure these measurands.

The main advantages of piezoelectric sensors are

- extremely high rigidity (measuring deflections are typically in the μm range),
- high natural frequency (up to over 500 kHz),
- extremely wide measuring range (span-to-threshold ratio up to over 10^8),
- very high stability (with transduction elements made of single crystals),
- high reproducibility,
- high linearity of the dependence of the output on the measurand,
- wide operating temperature range,
- insensitivity to electric and magnetic fields, and to radiation.

Table 1.1 Piezoelectric Sensors Versus Passive Sensors. The comparison is based on the stress sensitivity of the different systems. The figures give only a general indication to illustrate the key characteristics and orders of magnitude.

Transduction Principle	Strain Sensitivity V/$\mu\varepsilon$	Threshold (1 to 100 Hz) $\mu\varepsilon$	Span-to-threshold Ratio
Piezoelectric	5	0,00001	100000000
Capacitive	0,005	0,0001	750000
Inductive	0,001	0,00005	2000000
Piezoresistive (semiconductor) strain gage	0,0001	0,0001	2500000
Resistive (wire/metal film) strain gage	0,000005	0,01	50000

Table 1.1 schematically compares the piezoelectric system with resistive, capacitive and inductive systems. The only disadvantage of piezoelectric sensors is that they are inherently unable to measure statically over a longer period of time. The reason is that there are no materials of infinitely high insulation resistance and no vacuum tubes or semiconductors completely free of leakage currents – prerequisites for true static measuring with piezoelectric sensors.

Passive sensors do not have this limitation because the change in the electric property caused by the measurand will inherently remain as long the measurand acts with the same magnitude on the sensor and it can be detected over an unlimited period of time by the always necessary external power supply.

The inability of piezoelectric sensors to measure in a true static way has led to a widespread and still persisting prejudice that only "dynamic" measurements or measuring only "very short events" or "high-frequency phenomena" were possible with them. As explained in chapter 11, quasistatic measurements are perfectly well possible with piezoelectric sensors having single crystals as transduction elements.

Especially because of this capability the field of application for piezoelectric sensors has enormously expanded during the past 50 years. The unique characteristics of piezoelectric sensors mentioned above – some unmatched by any other type of sensor – often have brought first-time solutions to hitherto unsolved measuring tasks. Numerous applications described in this book bear witness that piezoelectric sensors still have a large potential for finding new applications and for serving as reliable sensors in the most demanding applications.

2 Background of Piezoelectric Sensors

2.1
Direct and Converse Piezoelectric Effect

Piezoelectricity is understood as a linear electromechanical interaction between the mechanical and the electrical state in crystals without a center of symmetry.

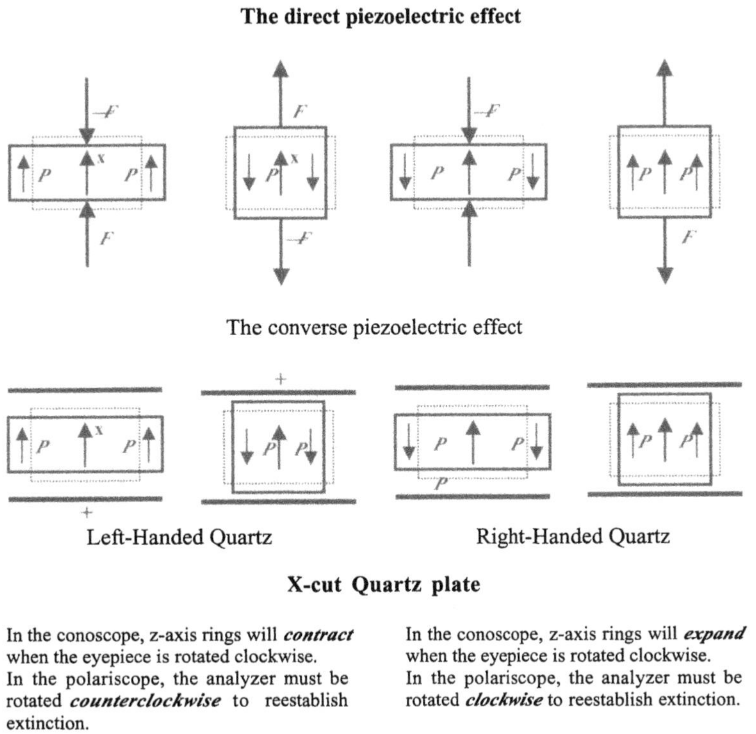

The direct piezoelectric effect

The converse piezoelectric effect

Left-Handed Quartz

Right-Handed Quartz

X-cut Quartz plate

In the conoscope, z-axis rings will ***contract*** when the eyepiece is rotated clockwise.
In the polariscope, the analyzer must be rotated ***counterclockwise*** to reestablish extinction.

In the conoscope, z-axis rings will ***expand*** when the eyepiece is rotated clockwise.
In the polariscope, the analyzer must be rotated ***clockwise*** to reestablish extinction.

Figure 2.1 Schematic representation of the piezoelectric effect in an X-quartz plate. Note: the polarities indicated follow IEEE Standards on Piezoelectricity (ANSI/IEEE Std 176-1987) [IEEE 1987]

The direct piezoelectric effect is present when a mechanical deformation of the piezoelectric material produces a proportional change in the electric polarization of that material, i.e. electric charge appears on certain opposite faces of the piezoelectric material when it is mechanically loaded. The converse piezoelectric effect means that mechanical stress proportional to an acting external electric field is induced in the piezoelectric material, i.e. the material is deformed when an electric voltage is applied.

The piezoelectric effect can best be illustrated with a quartz plate cut normal to the crystallographic x-axis (X-quartz plate) as shown in Fig. 2.1. The acting force $\boldsymbol{F}$ deforms the quartz plate and produces through the piezoelectric effect its electric polarization $\boldsymbol{P}$. In the converse piezoelectric effect the external electric field $\boldsymbol{E}$ between the electrodes induces the mechanical stress, which deforms the quartz plate.

2.2 Discovery of the Piezoelectric Effect

Since time immemorial, people in India and Ceylon – where tourmaline crystals are easily found – had observed that such crystals, when thrown into hot ashes, would strongly attract surrounding ash particles after a few moments, only to repel them again a little later. Dutch merchants brought this knowledge, together with the first tourmaline crystals, to Europe around 1703, where tourmaline was then often called "Ceylon magnet" [Dietrich 1985]. Linnaeus gave it the scientific name "lapis electricus" in 1747 already, although the electrical nature of this phenomenon was proven only in 1756 by Aepinus who noted the opposite polarities at the two ends of a heated tourmaline crystal. Brewster named this effect "pyrolectricity" (from the Greek "pyro", meaning fire) in 1824.

Already Haüy and Becquerel tried to find a relationship also between mechanical stress and electric polarization but the findings they reported were far from conclusive. The experimental setups at the time were always prone to be perturbed by the omnipresent triboelectricity (electric charge produced through friction) in the insulators used, such as amber.

The direct piezoelectric effect (from the Greek "piezin", meaning to press) was discovered by the brothers Pierre and Jacques Curie and first announced during the session of the Académie des Sciences in Paris on 2 August 1880 [Curie 1880]. They reported about an existing relationship between mechanical load and electric polarization but did not use the term piezoelectricity yet:

"Those crystals having one or more axes whose ends are unlike, that is to say hemihedral crystals with oblique faces, have the special physical property of giving rise to two electric poles of opposite signs at the extremities of these axes when they are subjected to a change in temperature: this is the phenomenon known under the name of pyroelectricity.

We have found a new method for the development of polar electricity in these same crystals, consisting in subjecting them to variations in pressure along their hemihedral axes. (...).

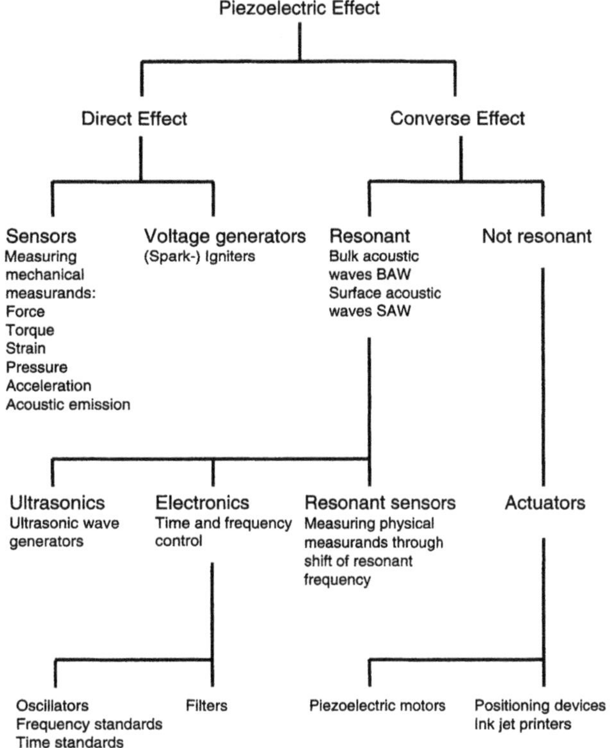

Figure 2.2 Applications of the piezoelectric effect

Our experiments were done with zinc blende, sodium chlorate, boracite, tourmaline, quartz, calamine, topaz, tartaric acid, cane sugar, and Rochelle salt. (...)."

The success of the two young physicists – Pierre was only 21 years old and Jacques just 3 years older – was by no means a coincidence. Rather it was the result of systematic studies of the symmetry in crystals and the effects observed in them. Such investigations characterize the first important phase of the development of solid state physics.

Lippmann predicted the existence of a converse piezoelectric effect based on thermodynamic considerations in 1881 [Lippmann 1881], saying that a crystal will deform when an electric voltage is applied to certain opposite faces of the crystal. The Curie brothers confirmed this in the same year, exhibiting remarkable experimental skills considering the technical means disposable at the time. The now commonly used term "piezoelectricity" was first proposed by Hankel [Hankel 1881] and immediately accepted in the field.

The work by the Curie brothers found a strong echo outside of France and prompted a number of famous physicists such as Röntgen, Kundt, Voigt [Voigt 1910] and Riecke [Riecke and Voigt 1892] to start their own investigations in the newly opened field.

However, for more than 30 years these effects remained a mere scientific curiosity without any practical application. Then, during WW1 when means of detecting submarines were being sought, Langevin conceived the idea of exciting quartz plates electrically into mechanical vibration through the converse piezoelectric effect for emitting high-frequency (ultrasonic) sound waves under water. The echo – ultrasonic waves bounced back by objects in the water or by the bottom of the sea – could be captured again by the same quartz plates, exploiting now the direct piezoelectric effect. This was the beginning of echo sounding, still indispensable today [Langeivn 1924; Langevin and Chilowsky 1926]. Langevin is rightfully considered the originator of the ultrasonic technology.

After WW1, it was also realized that materials exhibiting the direct piezoelectric effect could be used as transduction elements in sensors for measuring force, pressure and vibration (see section 2.4).

The field of application for the converse piezoelectric effect grew rapidly and today plays a key role in countless areas of modern life. Therefore, a brief outline is given in the next section to illustrate the not always obvious presence of these applications. The applications of the direct piezoelectric effect are still surprisingly little known and that is one of the major reasons for writing this book. Fig. 2.2 gives a general overview of the technical applications of the piezoelectric effect.

2.3 Applications of the Converse Piezoelectric Effect

The first attempts to generate ultrasound waves by exploiting the converse piezoelectric effect during the WW1 caught the curiosity of Cady who subsequently dedicated his whole life to the study of piezoelectricity. Through his countless publications which were crowned by a monograph published in 1964 [Cady 1964] and which is still an unsurpassed reference today, he earned the title of "father of modern piezoelectricity".

Already in 1918, Cady noticed peculiarities in the electrical behavior of Rochelle salt plates – to be used as source for underwater signaling – in the neighborhood of the frequencies of mechanical resonance of these plates. He so discovered the principle of the piezoelectric resonator and suggested in 1921 that it be used for stabilizing the frequencies of radio transmitters and for frequency filters [Cady 1921 and 1922]. The application of the converse piezoelectric effect in ultrasound and communication techniques is based on a simple principle. A piezoelectric element deforms periodically in an alternating electric field, constituting a electromechanical piezoelectric converter. If the frequency of this converter is tuned to the mechanical natural frequency of the piezoelectric element, it will start to resonate with a substantial amplitude. Since piezoelectric materials – above all quartz – have a very high Q-factor at their mechanical resonant frequency, this frequency is extremely well defined and stable.

Now the frequency range attainable with piezoelectrically generated ultrasonic waves (including the harmonics) is starting to get close to the THz range. The corresponding wave lengths are fractions of μm which makes numerous technical applications possible such as measuring the speed of sound, determining elastic material properties, echo sounding, materials testing, working materials, preparing extremely fine emulsions, eliminating dust and smoke particles, degassing or atomizing liquids, transmitting signals, remote controlling of equipment, medical diagnostic and therapy with ultrasound, and allows even to investigate molecular processes (molecular acoustics).

Considering the millions of quartz watches in use today – this also includes the even more numerous "hidden" piezoelectric timing circuits in devices such as radios, TV and their remote controls, video systems, computers, telephones, global positioning systems (GPS) and other navigation systems, chip-controlled household appliances, chip-controlled industrial control systems and so on – it is safe to assume that there are now probably more than 20 piezoelectric elements using the converse piezoelectric effect per household (at least in the industrialized countries). Also there is no other mass-produced measuring instrument which comes anywhere near the precision realized in quartz watches!

Another important application of the converse piezoelectric effect is in the field of actuators. Linear displacements in the nm to μm range can be produced with high accuracy and repeatability. Uses are e.g. positioning devices for mirrors in laser applications, micro-manipulators, linear motor drives, piezoelectric motors, actuators (e.g. in atomic force microscopy), ink jet printers and so on.

Probably the only application of the direct piezoelectric effect having found widespread use in low-cost mass products are the piezoelectric lighters and ingniters. More recently, piezoelectric switches and keypads have been gaining in importance, too. The direct piezoelectric effect is, however, the base of piezoelectric sensors – the topic of this book.

2.4 History of Piezoelectric Sensors

The first applications of the direct piezoelectric effect started to appear after WW1 when the first sensors for measuring force, pressure and acceleration (mainly in the form of shock and vibration) were designed, yet made for research purposes only, i.e. for specific experiments in the laboratory. Pressure and acceleration were the first measurands for which piezoelectric sensors were designed, while force sensors followed later.

The first publication on measuring pressure piezoelectrically was made in 1919 by Thomson [Thomson 1919] in England. Interest was focused on explosion pressures, a topic taken up by Hull [1921] and Keys [1921]. Karcher followed in 1922 [Karcher 1922] and in 1923, Tschappat described instruments specifically for measuring ballistic pressures [Tschappat 1923]. In 1925, Okochi in Japan was first to succeed in measuring the cylinder pressure in an internal combustion engine with a quartz pressure sensor of his own design at speeds up to over 3000 r/min [Okochi et al 1925].

The earliest description of a piezoelectric vibration sensor was published in England by Wood in 1921 [Wood 1921]. In 1928, Okochi et al described a balancing machine based on piezoelectric sensors [Okochi and Miyamoto 1928] and in 1929, Yamaguchi described an acceleration sensor, too [Yamaguchi 1929]. Zeller reported 1930 on a piezoelectric method to detect traffic vibrations [Zeller 1930]. In the same year, Kato et al published on the use of piezoelectric accelerometers for measuring the velocity of elastic waves produced by artificial shocks [Kato and Nakamura 1930]. Kishinouye described 1934 an experiment on the progression of fracture, recording the emitted sound with an inductive phonograph pickup directly on photographic film. He noted that "in addition to cracking sounds, inaudible vibrations were found to exist" [Kishinouye 1934]. These might well be the first attempts to measure acoustic emission (AE), although Förster and Scheil [Förster and Scheil 1936] are usually credited for the first instrumented experiment in 1936 specifically intended to measure AE (see 10.2). Kishinouye also studied the analogy between earthquake swarms and the failure mode of a material [Kishinouye 1937].

It was again Okochi who published the first cutting force measurements (in 3 components that still were resolved mechanically) made with quartz force sensors in 1927 [Okochi and Okoshi 1927] and in 1928, he used his quartz force sensor to measure the support load of a beam with a moving load [Okochi and Okoshi 1928]. In 1927, Watanabe described a new type of cathode ray oscillograph which served to indicate and measure the electric charge yielded by the sensor, and which replaced the very delicate string electrometers [Watanabe 1927]. The same year, Kluge reported on measuring "compression and acceleration forces" by piezoelectric methods [Kluge and Linckh 1929].

At this point of time – half a century after the discovery of the piezoelectric effect – the essential foundation was already established for the development of piezoelectric sensors, despite the rather limited number of publications on the subject (the listing above is nearly complete!). After 1930, the cadence of publications gained momentum and a very complete list of references up to around 1950 can be found in [Cady 1964, vol 2, pp 691-698 and 771-803]. A list of references up to about 1979 can be found in [Tichý and Gautschi 1980]. A short history of acceleration sensors was written by Walter [1997] and Drouillard [1990 and 1994] includes a large list of references on the history of AE.

A little known fact is that the French State Railways (SNCF) used quartz force sensors of their own design during the thirties and forties to measure the draw bar pull of locomotives in running trains. Most remarkable is that they measured this force for periods of up to several hours at a time, thus proving at that time already that it is indeed possible to measure quasistatically (see 11.5.5) with piezoelectric sensors. Unfortunately, the records and detailed description of these measurements were lost by the SNCF.

In the thirties, Zeiss-Ikon in Germany produced the first commercially available quartz sensors for force and pressure in small quantities. After WW2, several companies in Europe and the USA, in the seventies also in Japan and China, began to make piezoelectric sensors and the piezoelectric measuring principle started to find more and more practical applications.

In this historical review reference must also be made to a very particular type of piezoelectric "sensor" that is present in living creatures. In 1957, Fukuda and Yasuda were first to report that there exists a piezoelectric effect in bone [Fukuda and Yasuda 1957]. Although still not completely understood, it appears that the well known phenomenon that bone changes and adapts its internal structure continuously in such a way as to support the dominant load imposed on it with a minimum bone mass is indeed controlled by the piezoelectric effect. Examples are the reduction of bone mass in long-term bedridden patients and in astronauts. Bone acts like a piezoelectric force sensor and the electric charge yielded stimulates and controls the growth of bone as a function of the magnitude and direction of the forces acting on it. The piezoelectric effect – direct and converse – was found in various other biological materials, too. More references can be found e.g. in [Aschero et al 1996; and Taylor 1985].

3 Piezoelectric Materials for Sensors

Out of the increasing number of piezoelectric materials only a rather restricted number have been proven suitable for transduction elements in piezoelectric sensors. Basically, natural and synthetic single crystals, piezoelectric ceramics, textures and thin films can be used.

Here, only those materials which are used in commercially available sensors are introduced and their major characteristics described. More information on piezoelectric materials can be found in [Erhart et al 2002; Galassi 2000; Ikeda 1996; Lang 1990 and 1998; Philippot et al 2001; Wun-Fogle et al 1999; Xu 1991].

3.1 Requirements of Piezoelectric Materials for Sensors

Materials suitable for transduction elements in sensors ideally should have the following properties:

- high piezoelectric sensitivity,
- high mechanical strength,
- high rigidity (high modulus of elasticity),
- high electric insulation resistance (also at high temperatures),
- minimal hygroscopicity,
- linear relationship between mechanical stress and electric polarization,
- absence of hysteresis,
- high stability of all properties,
- low temperature dependence of all properties within a wide temperature range,
- low anisotropy of mechanical properties, such as thermal expansion coefficients and elastic constants,
- good machinability,
- low production cost.

The piezoelectric sensitivity is determined by the piezoelectric coefficients $d_{i\lambda}$. Their matrix depends on the crystal symmetry. The coefficient exploited in the sensor should be as large as possible while other coefficients usually only lead to undesirable side effects and hence should be as small and as few as possible.

Therefore materials with a low crystal symmetry, i.e. crystal cuts not normal to the crystal axes, are less suitable. Particular requirements must be met by the matrix of coefficients of materials to be used in multicomponent sensors (see 6.3).

A high mechanical strength is needed for measuring large measurands and to provide a good protection against overload. Also in miniaturized sensors the stress to which the material is subjected is usually very high.

A high rigidity is a sine qua non for attaining a high natural frequency and for minimizing the measuring deflection (see 5.2). This is obtained with small elastic coefficients $s_{\lambda\mu}$. A too pronounced anisotropy of the elastic properties in the surface plane on which the mechanical load acts leads to an anisotropy in the radial stress in the contact area with the load-applying elements which usually have isotropic elastic properties (e.g. steel). This creates a complex multiaxial state of stress which is detrimental to the piezoelectric sensitivity or even to the reproducibility in measuring large forces (see 3.2.4). Obviously a good machinability allows to produce the piezoelectric transduction elements more easily and at lower cost.

Only with a high insulating resistance it is possible to measure quasistatically with piezoelectric sensors (see 11.5.5). Besides the specific volume resistance the surface conductivity must be considered too, which depends heavily on the methods of working and treating the surface. The dependence of the insulation resistance on humidity and especially on temperature is a key factor in many applications such as in industrial environments and at high temperature. Often special treatment is needed to reduce the hygroscopicity of certain materials.

The long-term stability is essential for measuring reliably, a problem that can e.g. appear in α-quartz when simultaneous high temperature and high load may lead to twinning (see 3.2.4) or in piezoelectric ceramics suffering from aging effects (see 3.5.1). The study of the linear relationship between mechanical stress and electric polarization was only furthered by measuring the material constants of third order (see 3.2.7). For suppressing twinning and achieving a piezoelectric sensitivity as independent as possible of temperature and acting mechanical stress, not only the type of material but also the orientation of the cut element relative to the crystallographic axes must be considered.

The production cost is influenced by the cost of the material as well as by its machinability and a possibly required surface treatment.

3.2 Quartz

Quartz is still the most important single crystal serving as transduction element in piezoelectric sensors. Its chemical formula is SiO_2 and it is found in several modifications, constituted of oxygen tetrahedrons with silicon at the center position. For piezoelectric applications, the low-temperature modification of quartz existing below 573 °C is used. It is usually designated as α-quartz and belongs to the original-trapezoid class of symmetry 32 of the trigonal (original) crystal system. In each elementary cell there are 3 molecules of SiO_2. The threefold

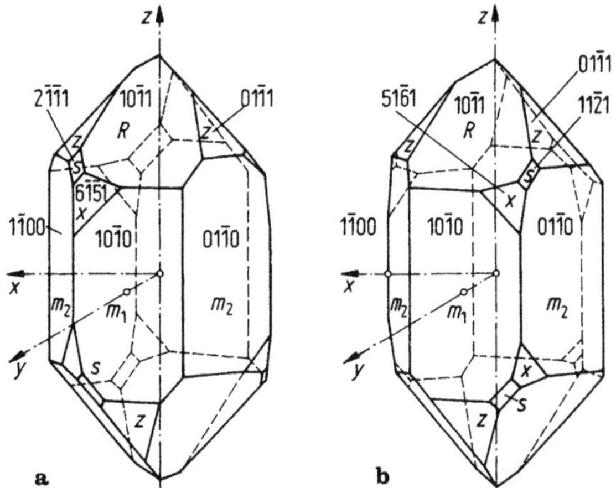

Figure 3.1 The 2 enantiomorph forms of α-quartz with Cartesian coordinate systems. **a** left-handed quartz, **b** right-handed quartz [Bechmann and Hearmon 1966]

axis of symmetry in the direction of the space diagonal is called the optic-crystallographic *c*-axis. Normal to it are 3 equivalent twofold crystallographic *a*-axes, also called electric axes in quartz (Fig. 3.1). The lattice constants are $a_0 = 0{,}491267\ nm$ and $c_0 = 0{,}540459\ nm$ at room temperature.

When the temperature is raised above 573 °C there is a phase change. The modification appearing belongs to the hexagonal-trapezoid class of symmetry 622 of the hexagonal crystal system. It is stable in the temperature range 573 ... 867 °C and usually designated as β-quartz or β-tridymite in the literature on piezoelectricity. However, some authors (e.g. [Sheludew 1975]) utilize exactly the opposite terms by denoting the low-temperature modification of quartz as β-quartz and the modification above 573 °C as α-quartz.

The phase change α-β-quartz is a phase order transition of the shifting type. It announces itself far below 573 °C by an increase in cell volume [Arnold 1976]. The atoms of the crystal structure shift continuously with increasing temperature in such a way that they are in locations of a higher central symmetry. Raman was first in observing a "soft oscillation" in the Raman spectrum during such a phase change [Raman 1940].

In the range 867 ... 1470 °C, α-tridymite and in the range 1470 ... 1723 °C, cristoballite is stable. Cristoballite is the most symmetric modification and has a diamond-like cubic structure.

3.2.1 Choice of Coordinate System

There are two enantiomorphic forms of α-quartz. They are called right-handed and left-handed quartz. The spatial group of right-handed quartz is $P3_221$, that of left-handed quartz $P3_121$. All processes in a right-handed quartz are the same as in a left-handed quartz, but in a mirror image. Both forms are represented schematically in Fig. 3.1. They can be distinguished by the position of the trapezohedral faces, by the etching patterns, by the reflection of X-rays on certain lattice planes and by the sense of rotation of the optical effect, from which the designation is derived. A right-handed quartz will rotate the plane of polarization of a light wave in the direction of the optical axis to the right. An observer looking at the light source will have to turn the analyzer clockwise to follow the rotation of the plane of polarization while a left-handed quartz rotates the plane of polarization in the opposite sense.

Formerly the choice of a Cartesian coordinate system in a right-handed quartz or a left-handed quartz was not standardized. Generally the x-axis was chosen in the direction of one of the 3 electric a-axes and the z-axis in the direction of the optical c-axis. Cady [Cady 1964] and some other authors [e.g. Petrzilka et al 1960] use a right-handed (positive) system for right-handed quartz and a left-handed (negative) system for left-handed quartz. Voigt [Voigt 1910] uses for both, right- and left-handed quartz, a uniform right-handed system with the axes oriented in such a way that c_{12}, d_{11}, e_{11}, and e_{14} take a positive value while s_{14} and d_{14} have a negative value. Based on the recommendation by the "IRE Standards on Piezoelectric Crystals, 1949" [IRE 1949 and 1958] a standardized right-handed coordinate system for right- and left-handed quartz has gained general acceptance.

Depending on the coordinate system chosen, some material constants will have different signs. The signs of all elastic constants are the same for left- and right-handed crystals. All piezoelectric constants have opposite signs for left- and right-handed crystals. Table 3.1 lists these material constants with their respective signs in the corresponding coordinate systems for right- and left-handed quartz.

For designating the orientation of the crystal cuts we choose the method used in piezoelectric resonators [IRE 1949 and 1958; Petrzilka et al 1960; N.N. 1957]. One assumes a piezoelectric element in the form of a rectangular solid with the length l, width b and thickness a (Fig. 3.2). The symbol attributed to a certain

Table 3.1 Signs of material constants that depend on the coordinate system used

Standard	IRE 1949		Cady		Voigt		IEEE 1978		Quartz	
Quartz	RH	LH	RH	LH	RH	LH	RH	LH	RH	LH
Coordinate system	RH	RH	RH	LH	RH	RH	RH	RH	RH	LH
s_{14}	+	+	–	–	–	–	–	–	–	–
c_{14}	–	–	+	+	+	+	+	+	+	+
d_{11}	–	+	+	+	+	–	+	–	+	–
d_{14}	–	+	–	–	–	+	–	+	–	+
e_{11}	–	+	+	+	+	–	–	+	+	–
e_{14}	+	–	+	+	+	–	–	+	+	–

LH = left-handed, RH = right-handed

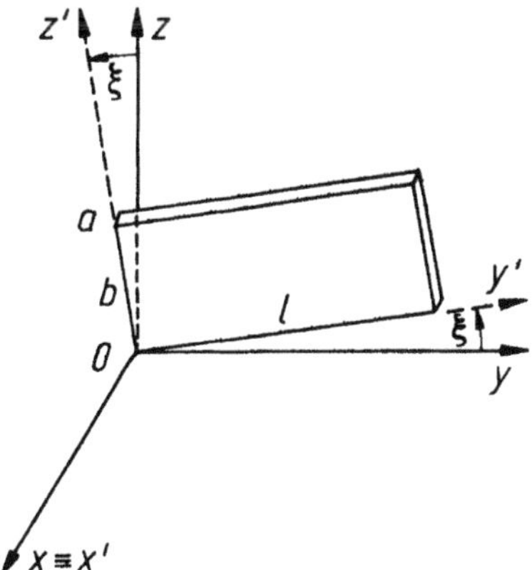

Figure 3.2 A crystal element with the orientation XYaζ

crystal cut indicates how it is derived through consecutive rotations about the edge of the rectangular solid from a reference state. The reference state is chosen in such a way that the edges of the rectangular solid are parallel to the axes x, y, and z of the Cartesian coordinate system. The first two letters of the symbol indicate, in sequence, the directions of the thickness and the length of the rectangular solid in its reference state. If several rotations are needed, they are indicated in the intended sequence. The second and third rotation refers to the position of the edges which were reached through the previous rotation. For a circular disc the direction of the thickness and of the diameter about which the plate must be rotated are indicated for the reference state. If no rotation is required, only the direction of the thickness needs to be specified. This way of designating crystal cuts is, of course, not limited to quartz only, but used generally.

3.2.2 Physical Properties

Quartz has a density of $\rho = 2{,}649 \cdot 10^3$ kg/m and a hardness of 7 on the Mohs' scale. It is practically insoluble in water and resistant against most acids and alkali. Quartz melts at 1710 °C.

For the application of α-quartz in piezoelectric sensors its mechanical strength is of prime importance. The resistance to compression of quartz plates is, depending on their orientation, in a range of about 2 ... 3 GPa. Berndt [Gmelin 1959, p 329] reports a maximum strength of quartz cylinders when compressing them between two steel plates of 2,75 GPa when the axis of the cylinder is parallel to the optical axis c and of 2,7 GPa when the axis of the cylinder is normal to c, whereby the compression force is acting in the direction of the axis of the cylinder. Under hydrostatic pressure the strength of quartz is even higher and reaches about

Table 3.2 Physical properties of a-quartz (for left handed quartz)

Matrix of elastic coefficients

$$\begin{pmatrix} s_{11} & s_{12} & s_{13} & s_{14} & 0 & 0 \\ s_{12} & s_{11} & s_{13} & -s_{14} & 0 & 0 \\ s_{13} & s_{13} & s_{33} & 0 & 0 & 0 \\ s_{14} & -s_{14} & 0 & s_{44} & 0 & 0 \\ 0 & 0 & 0 & 0 & s_{44} & 2s_{14} \\ 0 & 0 & 0 & 0 & 2s_{14} & 2(s_{11}-s_{12}) \end{pmatrix}$$

Matrix of elastic moduli

$$\begin{pmatrix} c_{11} & c_{12} & c_{13} & c_{14} & 0 & 0 \\ c_{12} & c_{11} & c_{13} & -c_{14} & 0 & 0 \\ c_{13} & c_{13} & c_{33} & 0 & 0 & 0 \\ c_{14} & -c_{14} & 0 & c_{44} & 0 & 0 \\ 0 & 0 & 0 & 0 & c_{44} & c_{14} \\ 0 & 0 & 0 & 0 & c_{14} & \frac{1}{2}(c_{11}-c_{12}) \end{pmatrix}$$

Elastic coefficients in $10^{-12}\,N^{-1}m^2$

		s_{11}^E	s_{12}^E	s_{13}^E	s_{14}^E	s_{33}^E	s_{44}^E	$[s_{66}^E]$
Adiabatic	(25°C)	12,777	–1,807	–1,235	4,521	9,735	19,985	29,167
Isothermal	(25°C)	12,809	–1,775	–1,218	4,521	9,743	19,985	29,167

Temperature coefficients of elastic coefficients in $10^{-6}K^{-1}$ (25°C)

$TK(s_{11}^E)$	$TK(s_{12}^E)$	$TK(s_{13}^E)$	$TK(s_{14}^E)$	$TK(s_{33}^E)$	$TK(s_{44}^E)$	$[TK(s_{66}^E)]$
8,5	–1296,5	–168,8	140,6	139,7	211,1	–151,9

Elastic moduli in $10^9\,Nm^{-2}$

		c_{11}^E	c_{12}^E	c_{13}^E	c_{14}^E	c_{33}^E	c_{44}^E	$[c_{66}^E]$
Adiabatic	(25°C)	86,80	7,04	11,91	–18,04	105,75	58,20	39,88
Isothermal	(25°C)	86,48	6,72	11,66	–18,04	105,55	58,20	39,88

Temperature coefficients of elastic moduli in $10^{-6}K^{-1}$ (25°C)

$TK(c_{11}^E)$	$TK(c_{12}^E)$	$TK(c_{13}^E)$	$TK(c_{14}^E)$	$TK(c_{33}^E)$	$TK(c_{44}^E)$	$[TK(c_{66}^E)]$
–44,3	–2690	–550	117	–160	–175,4	187,6

Matrix of piezoelectric coefficients

$$\begin{pmatrix} d_{11} & -d_{11} & 0 & d_{14} & 0 & 0 \\ 0 & 0 & 0 & 0 & -d_{14} & -2d_{11} \\ 0 & 0 & 0 & 0 & 0 & 0 \end{pmatrix}$$

Matrix of piezoelectric moduli

$$\begin{pmatrix} e_{11} & -e_{11} & 0 & e_{14} & 0 & 0 \\ 0 & 0 & 0 & 0 & -e_{14} & -e_{11} \\ 0 & 0 & 0 & 0 & 0 & 0 \end{pmatrix}$$

Table 3.2 continued

Piezoelectric coefficients (20 °C) in $10^{-12}CN^{-1}$		in $C^{-1}m^2$		Piezoelectric moduli (20 °C) in Cm^{-2}		in 10^9NC^{-1}	
d_{11}	d_{14}	g_{11}	g_{14}	e_{11}	e_{14}	h_{11}	h_{14}
2,30	0,67	0,0578	0,0182	0,171	–0,041	4,36	–1,04

Temperature coefficients of piezoelectric coefficients in $10^{-4}K^{-1}$ (20 °C)		Temperature coefficients of piezoelectric moduli in $10^{-4}K^{-1}$ (20 °C)	
$TK(d_{11})$	$TK(d_{14})$	$TK(e_{11})$	$TK(e_{14})$
–2,15	12,9	–1,6	–14,4

Matrix of permittivities	Relative permittivities (20 °C)				Temperature coefficients of relative permittivities $10^{-4}K^{-1}$ (20 °C)		Thermal expansion coefficients in $10^{-6}K^{-1}$ (20 °C)	
$\begin{pmatrix} \varepsilon_{11} & 0 & 0 \\ 0 & \varepsilon_{11} & 0 \\ 0 & 0 & \varepsilon_{33} \end{pmatrix}$	$\left(\frac{\varepsilon_{11}}{\varepsilon_0}\right)^T$	$\left(\frac{\varepsilon_{33}}{\varepsilon_0}\right)^T$	$\left(\frac{\varepsilon_{11}}{\varepsilon_0}\right)^T$	$\left(\frac{\varepsilon_{33}}{\varepsilon_0}\right)^T$	$TK\left(\frac{\varepsilon_{11}}{\varepsilon_0}\right)$	$TK\left(\frac{\varepsilon_{33}}{\varepsilon_0}\right)$	α_{11}	α_{33}
	4,514	4,634	4,428	4,634	0,5	0,5	13,71	7,48

4 GPa according to Bridgman [Bridgman 1941]. Tensile strength is much lower, however, and is about 120 MPa in quartz rods with their axes in the direction of the optical axis *c* and about 97 MPa with the rod axis normal to *c*.

The coefficients of thermal expansion in α-quartz at room temperature are $\alpha_{11} = 13{,}7 \cdot 10^{6}/\,°C$ and $\alpha_{33} = 7{,}4 \cdot 10^{6}/\,°C$. They increase with rising temperature. In the temperature range 0 ... 400 °C, the average coefficients of thermal expansion are $\alpha_{11} = 18{,}5 \cdot 10^{6}/\,°C$ and $\alpha_{33} = 10{,}5 \cdot 10^{6}/\,°C$ (Fig. 3.3).

α-quartz is an excellent electrical insulator. The specific resistance of natural quartz at room temperature in the direction of the optical axis is $\varrho_c = 5 \ldots 20$ TΩm and for synthetic quartz $\varrho_c = 1 \ldots 10$ PΩm. In the direction normal to the optical axis the specific resistance is higher by at least two orders of magnitude. It drops substantially with increasing temperature and this relationship can be described as an exponential dependence on the reciprocal value of the absolute temperature [Kolodieva and Firsova 1969]. For piezoelectric sensors the surface conductivity is of particular importance [Tschapek et al 1969].

The elastic, piezoelectric and dielectric properties of quartz are summarized in table 3.2 which also indicates the temperature coefficients of the elastic and piezoelectric coefficients as well as of the moduli. These values were collected from

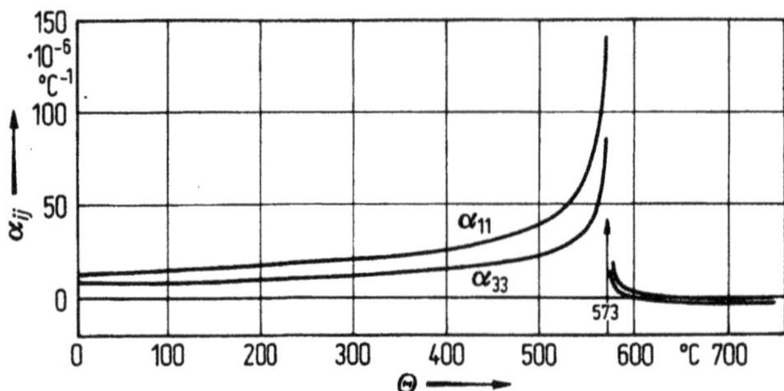

Figure 3.3 Temperature dependence of the expansion coefficients of quartz [Mayer 1959]

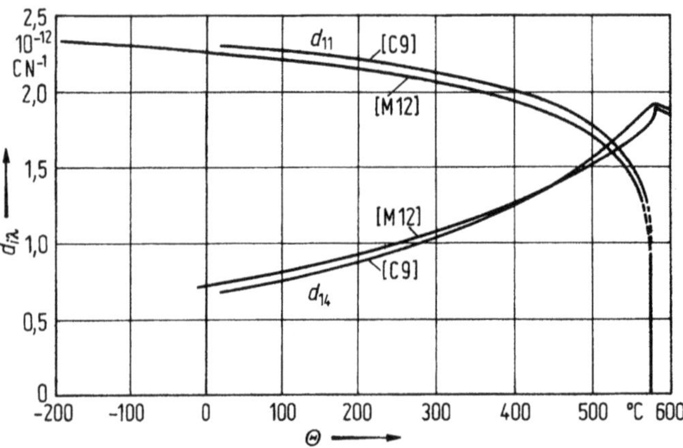

Figure 3.4 Temperature dependence of the piezoelectric coefficients of α-quartz (for left-handed quartz) [Mayer 1959; Cook 1950]

[Bechmann 1953 and 1962; McSkimin 1965; Thurston 1966; Thurston and Brugge 1964; Zelenka and Lee 1971]. The temperature dependence of the piezoelectric coefficients is also shown on Fig. 3.4, because it is particularly important in quartz elements used in piezoelectric sensors. More data can be found in [Brice 1985].

3.2.3 Synthetic Quartz Crystals

Quartz is, after feldspar, the most common mineral found on earth. However natural quartz crystals of sufficient size and quality suitable for use in technical applications of the piezoelectric effect are found mainly in Brazil. The growing demand for quartz crystals to be used as piezoelectric resonators and their increasing importance during World War II and the years after the war not only accelerated the search for substitute materials but also lead to the production of synthetic quartz by means of the hydrothermal synthesis [Liebertz 1973]. It can be done in thick-walled steel autoclaves at a pressure between 1 and 2 kbar and at temperatures between 350 and 400 °C. Water with a small addition of Na_2CO_3 or

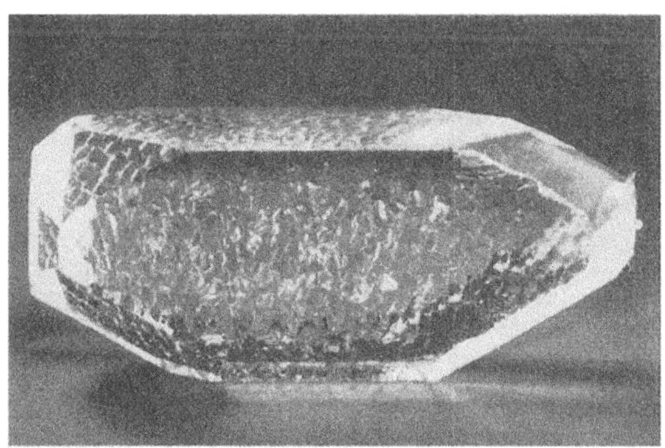

Figure 3.5

Figure 3.6

Figure 3.5 An artificially grown quartz crystal (Courtesy of Kistler)
Figure 3.6 Quartz crystals coming out of the autoclave (Courtesy of Salford Electrical Instruments)

NaOH (0,5 to 1 mol/l) serves as a solvent. The transfer of material is mainly by convection. Large quartz crystals (Figs. 3.5 and 3.6) with a mass of over 1 kg may take several weeks to grow.

The techniques for working quartz is described extensively by Heising [1946].

3.2.4 Twinning

In α-quartz so-called Brazilian, Dauphiné and Japanese twins occur because of the enantiomorphy. The Brazilian twins are formed by a right- and a left-handed quartz which are intergrown in such a way that the faces of both rhomboids (1120) are congruent in several positions while the optical axes *c* remain parallel. The plane of polarization of a light beam in the direction of the optical axis *c* [0001] is rotated in opposite directions by the twins. Therefore Brazilian twins are also called optical twins.

In Dauphiné twins 2 right-handed or 2 left-handed quartzes are rotated against each other by 60° and intergrown in such a way that the optical axes are parallel but the electric axes point in opposite directions (Fig. 3.7). Dauphiné twins are therefore called electrical twins, too.

In Japanese twins the optical axes form an angle of 84° 33'. Japanese twins can be composed of either a) a right- and a left-handed quartz, or b) of 2 right-handed quartzes, or c) of 2 left-handed quartzes.

Brazilian and Japanese twins of type a (Fig. 3.7) are mirror twins. It is impossible to bring the crystal lattices of the two individuals to congruency by a rotation which is the reason why they can not be formed by a mechanical or any other external influence.

In Dauphiné twins, however, the crystal lattice of both individuals are only rotated against each other. Such twins can form in a not twinned α-quartz by external influences, too. This is then called a secondary twinning.

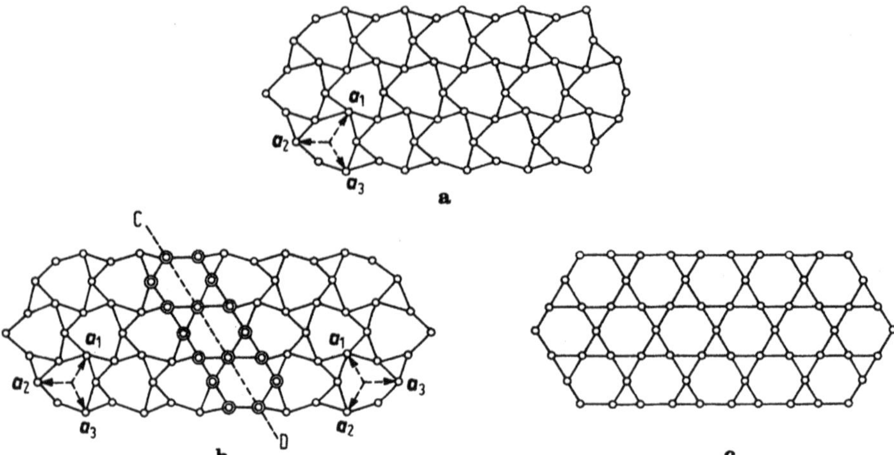

Figure 3.7 **a** Simplified representation of the projection of the Si-atoms in the crystal lattice of α-quartz on the plane (0001). **b** Dauphiné twins; at the left of the twinning line CD is the reference state, at the right the structure is rotated by 60° about the optical axis. **c** Lattice structure of β-quartz [Klassen-Neklyudova 1964]

Also, in Japanese twins of types b and c (Fig. 3.7), the two individuals can be brought to congruency by rotation about an axis normal to the plane of intergrowing. Theoretical considerations [Klassen-Neklyudova 1964] indicate that a force acting in the plane (1122) could lead to the formation of secondary twins. However, in such experiments only the formation of Dauphiné twins could be observed so far.

Secondary twinning does not lead to a change in form of the crystal [Klassen-Neklyudova 1964]. It occurs only if it is associated with a reduction of Gibbs' free energy. In the thermodynamic equilibrium we denote with $G(1)$ the thermodynamic Gibbs' free energy per unit of volume in the reference state and in the twinned state (after the reorientation of the a-axis) with $G(2)$. The formation of the twins changes the sign of the coefficients, which are framed in the $c_{i\lambda}$ and $s_{\lambda\mu}$ matrices:

$$\begin{pmatrix} d_{11} & -d_{11} & 0 & d_{14} & 0 & 0 \\ 0 & 0 & 0 & 0 & -d_{14} & -2d_{11} \\ 0 & 0 & 0 & 0 & 0 & 0 \end{pmatrix}, \tag{3.1}$$

$$\begin{pmatrix} s_{11} & s_{12} & s_{13} & s_{14} & 0 & 0 \\ s_{12} & s_{11} & s_{13} & -s_{14} & 0 & 0 \\ s_{13} & s_{13} & s_{33} & 0 & 0 & 0 \\ s_{14} & -s_{14} & 0 & s_{44} & 0 & 0 \\ 0 & 0 & 0 & 0 & s_{44} & 2s_{14} \\ 0 & 0 & 0 & 0 & 2s_{14} & 2(s_{11}-s_{12}) \end{pmatrix} \tag{3.2}$$

The thermodynamic condition for the formation of twins in α-quartz, considering its symmetry, is:

$$\begin{aligned} \Delta G &= G(2) - G(1) = \\ &= 2s_{14}\left(T_1 T_4 - T_2 T_4 + 2T_5 T_6\right) + 5d_{11}\left(E_1 T_1 - E_1 T_2 - 2E_2 T_6\right) < 0 \ . \end{aligned} \tag{3.3}$$

The development of the formulae leading to the equation for ΔG can be found in [Erhart et al 2002] in the chapter on ferroics.

Therefore, according to the terminology of Aizu [1970 and 1973] and Newnham [Newnham 1974 and 9175; Newnham and Cross 1974], α-quartz belongs to the class of ferrobielastics and potentially to the class of ferroelastoelectrics, too. However, its ferroelastoelectric properties do not seem to have been demonstrated experimentally so far.

One of the most frequently used type of quartz element in piezoelectric sensors is the X-quartz plate. Assuming a uniaxial state of stress in the direction of the x-axis (only $T_1 \neq 0$) twinning in such a plate would not entail a change in the Gibbs' free energy and therefore one would not expect any twinning. However when compressing such quartz plates and also in sensors twinning can be observed to start at room temperature when the stress reaches about 500 ... 900 MPa (5 ... 9 kbar). The reason for this is the multiaxial state of stress caused by bending of the load-introducing steel plates and by the different lateral expansions of quartz (here, lateral expansion is anisotropic as well) and steel. With increasing temperature twinning will already start at progressively lower stress levels and close to the temperature of the phase change at 573 °C, twinning can even be observed in the unloaded state of the crystal.

The presence of Dauphiné twins can be visualized by etching patterns (Figs. 3.8 and 3.9) obtained after etching the quartz crystals with e.g. hydrofluoric acid [Cady 1964, p 419; Heising 1946, p 164; Joshi and Kotru 1968; Joshi and Vag 1968] due to the different intensities of reflection on the lattice plane $(h\ k\ l)$ and $(h\ k\ l)$ [Arnold 1976; Inoue et al 1974; McLaren 1969]. Twinning is accompanied by a sudden change in the electric polarization which can clearly be seen when plotting

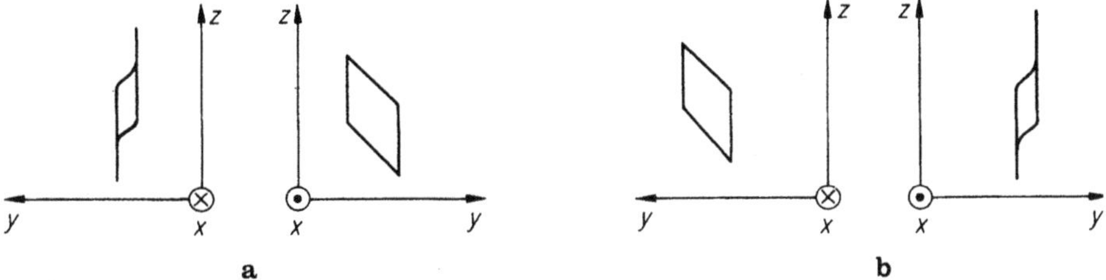

Figure 3.8 Polarization charges under compressive loading in the direction of the x-axis and schematized etching patterns (viewed from above) in an X-quartz plate from a right-handed quartz, relative to the coordinate system assigned to the reference state.
a reference state, **b** twinned state

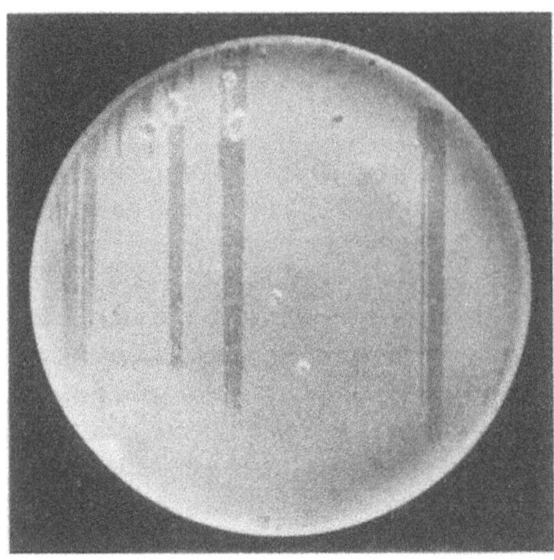

Figure 3.9 Dauphiné twins in an X-quartz plate, visible after etching and under oblique illumination (Courtesy of Kistler)

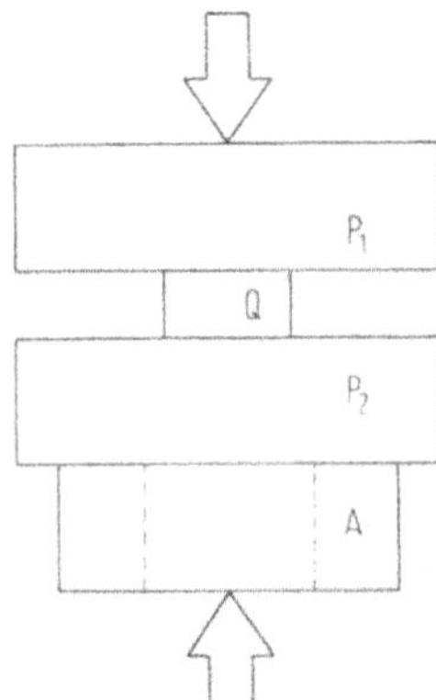

Figure 3.10 Schematic drawing of the experimental setup for investigating the formation of twins. Q quartz element; P_1, P_2 pressure plates; A piezoelectric sensor for measuring the applied force

the dependence of the electric flux density D against the mechanical stress T, as Fig. 3.11 shows. Twinning has also been observed directly in polarized light [Aizu 1972 and 1973; Dolino 1973] and by using the Schlieren method [Bertagnolli 1979; Bertagnolli et al 1978].

Twins forming during mechanical loading have been observed to completely disappear again when the load was taken off. Such vanishing of twins in an X-quartz plate that is compressed in the direction of its thickness (Fig. 3.10) can also be seen in the schematized dependence of D_1 on T_1 shown in Fig. 3.11 a. Up to point A the curve follows the equation $D_1 = d^*_{11}(0)\, T_1$. The proportionality factor $d^*_{11}(0)$ is the effective piezoelectric coefficient, which takes in account the lateral expansion imposed by the steel plates. In point A, twinning starts and the reorientation of the twinned areas under mechanical load means that their polarity is reversed. This causes a sudden reduction of D_1 up to point B. The twin so formed remains stable under further increasing load between points B and C and leads to a reduced

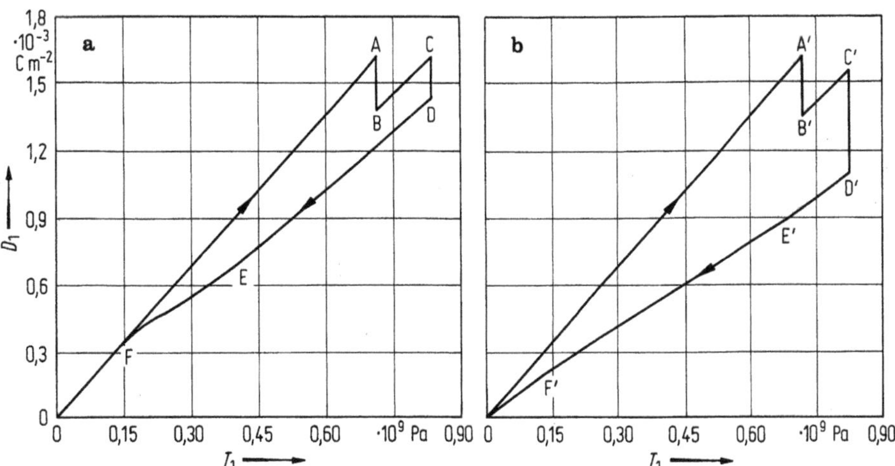

Figure 3.11 Schematic graph of the dependence of the piezoelectric flux density D_1 on the mechanical stress T_1. **a** complete reversal of the twins, **b** partial reversal of the stable twins

piezoelectric sensitivity $d_{11}^*(B)$. Therefore the slope of the straight line BC is smaller than that of section 0A. Between points C and D, twinning occurs again, accompanied again by a sudden reduction D_1, and leading to a further reduction of the piezoelectric sensitivity to the value of $d_{11}^*(D)$, whereby $d_{11}^*(D) < d_{11}^*(B) < d_{11}^*(0)$.

During unloading D_{11}^* decreases proportionally with T_1 and the slope of the straight line DE is determined by $d_{11}^*(D)$. In point E a new phenomenon is observed: because of the gradual vanishing of the twins the decrease of D_1 is slowed and at the same time the piezoelectric sensitivity increases again. In point F all twins have disappeared and the piezoelectric sensitivity returns to its original value $d_{11}^*(0)$. Below point F the quartz plate behaves again as it did before the formation of twins.

The twinning between points A and B or C and D depends on the duration of the mechanical load. If the load persists long enough, twinning may occur over larger areas and "stable" twins may be formed. Instead of the dependence of D_1 on T_1 as illustrated in Fig. 3.11 a, the relation represented schematically in Fig. 3.11 b is observed. The reversing of the twinning process between points E' and F' occurs over a broader range of T_1 and point F' remains below the ascending straight line 0A'. This signifies that not all twins have disappeared during the unloading. The remaining "stable" twins provoke a reduction of the piezoelectric sensitivity of the quartz plate and they can easily be visualized by etching the plate. The formation of "stable" twins can be favored not only by the duration of the mechanical load but also by a higher magnitude of the load and by elevated temperature.

As already mentioned we must assume that the stresses T_2, T_4, T_5 and T_6 developing in the contact area – a multiaxial stress state – are responsible for the formation of twins in an X-quartz plate that is compressed in the direction of its x-axis. The mechanism of friction between the steel plate and the quartz plate, which can even cause local plastic deformations, may result in slightly different multiaxial stress states during loading and unloading with the same force. The stress state can fulfill the thermodynamic conditions for twinning during loading and for reversal of the twinning during unloading.

Optical observation shows that twinning usually starts at points of highest stress concentration, in the contact surface with the steel plate or at points of defects in the quartz plate. The twins grow first in the direction of the crystallographic x-axis and expand later in the direction of the z-axis. The borderlines of the twinned areas are usually straight at room temperature but may become very irregular at higher temperature [Bertagnolli et al 1977].

3.2.5 Suppressing Secondary Twinning

A uniaxial stress state oriented randomly relative to the crystallographic axes can favor ($\Delta G<0$) or hinder ($\Delta G>0$) twinning. For expressing mathematically the necessary thermodynamic conditions, we choose a new Cartesian coordinate system $(0, x'_1, x'_2, x'_3)$ which is rotated against the crystal coordinate system $(0, x, y, z) \equiv (0, x_1, x_2, x_3)$. This is derived from the crystal coordinate system by a first rotation about the x_3-axis through an angle of ζ and then through a rotation about the x'_1-axis determined by the first rotation through an angle of ξ. Let us assume that there is only one stress component T'_2 in the direction of the x'_2-axis and no electric field present.

The change in the density of the Gibbs' free energy through twinning is

$$\Delta G = \frac{1}{2}\left(s'_{22}(1) - s'_{22}(2)\right)T'^{2}_{2}. \tag{3.4}$$

The transformation matrix for the described double rotation of the coordinate system is

$$\begin{pmatrix} \cos\zeta & \sin\zeta & 0 \\ -\sin\zeta\cos\xi & \cos\zeta\cos\xi & \sin\xi \\ \sin\zeta\sin\xi & -\cos\zeta\sin\xi & \cos\xi \end{pmatrix} \tag{3.5}$$

With the aid of this equation we obtain the transformation equations for $s'_{22}(1)$ and $s'_{22}(2)$. They differ from each other only by the signs of the terms which relate to the elastic coefficients framed in the matrix (3.2) and we, therefore, obtain

$$\Delta s'_{22} = s'_{22}(1) - s'_{22}(2) = -4\cos 3\zeta \sin\xi \cos^3\xi s_{14} \tag{3.6}$$

and

$$\Delta G = -2\cos 3\zeta \sin\xi \cos^3\xi s_{14} T'^{2}_{2}. \tag{3.7}$$

The value of the function $\Omega(\zeta,\xi)=\cos 3\zeta \sin\xi \cos^3\xi$ determines the formation of twins. Considering that in quartz $s_{14}>0$ (see [Beachman and Hearmon 1966]), twinning will be prevented if $\Omega<0$. The value of Ω is a measure for the tendency to twin in a uniaxial stress state. The highest resistance against twinning can be expected when

$$\Omega=\min. \tag{3.8}$$

Positive values of Ω favor twinning. The dependence of the inclination towards twinning on the stress state can also be exploited to remove twins, especially in natural quartz [Thomas and Rycroft 1946; Thomas and Wooster 1951; Wooster and Wooster 1946; Wooster et al 1947].

The condition for maximum resistance against twinning is fulfilled e.g. by a quartz rod with the orientation XYa 150°, cut for the transverse piezoelectric effect. Loading will be in the direction of the rotated long axis $y'=x'_2$ and the electrodes cover the faces normal to the $x=x_1$-axis. In quartz elements cut for the longitudinal piezoelectric effect it is impossible to fulfill the condition (3.8) because the piezoelectric sensitivity for such orientations is zero. Therefore one has to accept a compromise between an adequately high resistance against twinning and a still sufficient piezoelectric sensitivity required for practical applications.

Although the suppression of twinning through the choice of a suitable orientation of the crystallographic axes of the quartz element has been proven experimentally [Bertangnolli et al 1977; Calderara et al 1971] it has severe drawbacks. The transition to the rotated coordinate system assigned to the quartz element increases the number of elements in the transformed matrix of the piezoelectric coefficients that are not zero. This means that compared with an X-quartz plate additional mechanical stresses will contribute to the piezoelectric polarization of such a quartz element. Apart from the piezoelectric effect used for measuring such an element exhibits undesirable shear and transverse sensitivities, which have to be compensated for by taking special measures.

For piezoelectric sensors it would be more advantageous to suppress the twinning in an X-quartz plate or at least to increase the threshold of mechanical loading at which twinning starts. Thermodynamic considerations offer – at least theoretically – two possibilities: either one succeeds through radial preload in a suitable direction to achieve that for negative values of T_1 the change in Gibbs' free energy is positive, i.e. $\Delta G>0$; or it is possible to reduce the mechanical stresses responsible for twinning by choosing a material of suitable properties. Neither of these two ideas have yet been realized in practice.

3.2.6 Temperature Dependence of Piezoelectric Constants

Fig. 3.4 shows the temperature dependence of the piezoelectric coefficient d_{11}. The piezoelectric sensitivity of an X-quartz plate and an XY-cut for the transverse effect (in quartz, $d_{12}=-d_{11}$) drops considerably with increasing temperature. The value of the second, independent piezoelectric coefficient d_{14} increases with higher temperature. This makes it possible to find a quartz cut for the transverse effect in which the 2 temperature dependencies compensate each other and in which the effective piezoelectric coefficient remains temperature-independent at a certain temperature (in practice, in a certain temperature range).

Determining such a crystal cut is easy. Based on the formulae developed in (Erhart et al 2002) we can calculate the piezoelectric coefficient of a quartz cut XYaξ for the transverse effect whose long axes y' is rotated about the x-axis through the angle ξ against the crystallographic y-axis. We obtain

$$d'_{12} = -d_{11}\cos^2\xi + d_{14}\sin\xi\cos\xi \,. \tag{3.9}$$

Setting the first derivative of d'_{12} with respect to the temperature Θ equal to zero, we obtain for ξ the condition

$$\tan\xi = \frac{\dfrac{\partial d_{11}}{\partial \Theta}}{\dfrac{\partial d_{14}}{\partial \Theta}} \,. \tag{3.10}$$

To be able to conveniently mark the temperature dependence of the piezoelectric coefficient $d_{i\lambda}$, we define its temperature coefficient $TK(d_{i\lambda})$ by the relation

$$TK(d_{i\lambda}) = \frac{1}{d_{i\lambda}}\frac{\partial d_{i\lambda}}{\partial \Theta} \,. \tag{3.11}$$

This allows to write (3.10) in the form of

$$\tan\xi = \frac{d_{11}TK(d_{11})}{d_{14}TK(d_{14})} \,. \tag{3.12}$$

Because the temperature coefficients $TK(d_{11})$ and $TK(d_{14})$ are in turn temperature-dependent too, the condition $TK(d'_{12})=0$ can only be satisfied at one temperature for a chosen angle ξ. From the average temperature coefficients $TK(d_{11})$ and $TK(d_{14})$ for the temperature range 0...400 °C we obtain $\xi \approx 155°$ and the corresponding piezoelectric coefficient is $d'_{12}=-2{,}15$ pC/N (for left-handed quartz), i.e. about 93% of the value of d_{11}.

Our analysis is however valid only for an idealized model of a quartz element under just a uniaxial stress state in the direction of its long axis. In reality the interactions between the load-introducing steel plates and the quartz plate lead to a multiaxial stress state (a partially clamped state) and the optimal value for ξ becomes a little higher for obtaining the smallest possible temperature dependence of the piezoelectric sensitivity of the quartz element. Assuming that only the deformation in the long axis of the rod is not zero ($S_2 \neq 0$), we obtain instead of (3.12)

$$\tan\zeta = \frac{e_{11} TK(e_{11})}{2e_{14} TK(e_{14})} \ . \tag{3.13}$$

This condition yields a value of $\xi \approx 165°$, which may be considered as an upper limit for the orientation of a quartz element having the smallest possible temperature dependence of the piezoelectric sensitivity of the transverse effect. At the same time, such quartz elements have an optimal resistance against twinning (see 3.7 and [Calderara 1971; Calderara et al 1971]). In principle it is impossible to reduce the temperature dependence of the piezoelectric sensitivity in the same way for the piezoelectric longitudinal effect, because it is only a function of d_{11}. A small improvement can be obtained by partially clamping the quartz element.

3.2.7
Nonlinear Electromechanical Properties of α-Quartz

Quartz was one of the first crystals in which the electrooptic effect was discovered. Its electrooptic constants are $\varepsilon_{111} = 2{,}2 \cdot 10^{-23}$ F/V and $\varepsilon_{231} = 5{,}2 \cdot 10^{-23}$ F/V, and $r_{11} = 4{,}7 \cdot 10^{-13}$ m/V and $r_{41} = 1{,}9 \cdot 10^{-13}$ m/V.

The electrostriction in α-quartz is small. It can be proven most easily in the direction of the optical *c*-axis [Bohaty 1975] because the electric field strength E_3 provokes only an electrostricitve deformation. The electrostricitive effect can be interpreted as follows [Erhart et al 2002]: For $\boldsymbol{E} = 0$, in α-quartz, $d_{33} = 0$. However the electric field E_3 induces the piezoelectric coefficient d_{33} which is proportional to E_3. The coefficient q_{333} of electrostriction is found by measuring the dependence of the converse piezoelectric effect provoked by E_3 on E_3. We obtain $q_{333} = 3 \cdot 10^{-23}\,\text{m}^2/\text{V}^2$. In order to obtain in a Z-cut of α-quartz the same "piezoelectric effect" as in an X-cut, a field of a magnitude of about 10^9 V/m would have to be applied. The electrostricitive effect is much stronger in ferroelelectric crystals.

The piezooptic constants of α-quartz are summarized in table 3.3. The values for the electroelastic coefficients in α-quartz, also indicated in that table, were

Table 3.3 Piezooptic and electroelastic constants of α-quartz

Piezooptic constants (of left-handed quartz per [Cady 1964])

$\Pi_{\lambda\mu}$ in $10^{-12}\,\text{m}^2\text{N}^{-1}$

Π_{11}	Π_{12}	Π_{13}	Π_{14}	Π_{31}	Π_{33}	Π_{44}	Π_{41}
1,110	2,500	1,970	–0,097	2,770	0,183	–1,015	–0,320

Electroelastic constants (of left-handed quartz per [Hruska 1977; Hruska and Kazda 1968; Hruska and Khogali 1971])

$g_{i\lambda\mu}$ in $10^{-23}\,\text{Cm}^2\text{N}^{-2}$

g_{111}	g_{131}	g_{141}	g_{122}	g_{124}	g_{134}	g_{144}	g_{315}
–62,3	–94,8	–151	5,4	27,7	26,5	–210,0	11,4

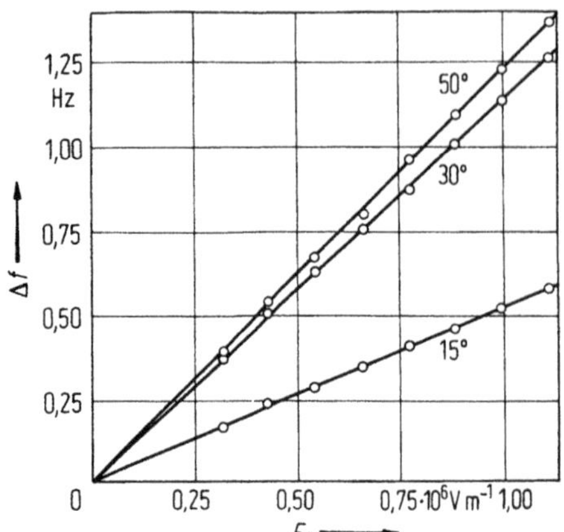

Figure 3.12 Dependence of the shift Δf in the resonant frequency on the the electric field strength E_1 in rod-shaped quartz resonators of the orientation XYaζ with $\zeta = 15°, 30°$ and $50°$ [Kinigadner et al 1977]

determined by evaluating measurements of the dependence of the resonant frequency of the longitudinal strain oscillations of rod-shaped quartz resonators in an electric field (Fig. 3.12) and similar measurements on resonators of suitable orientation, having other oscillating modes, too [Hruska 1971 and 1977; Hruska and Kazda 1968; Hruska and Khogali 1969 and 1971; Kinigadner et al 1977].

Thurston, McSkimin and Andreatch [Thurston et al 1966] investigated the dependence of the ultrasonic speed on the mechanical stress in α-quartz and determined its 14 independent moduli of elasticity of third order at 25 °C. For an X-quartz plate, the value of $c_{111} = (-2{,}10 \pm 0{,}07) \cdot 10^{11}\,\mathrm{N/m^2}$ is of special importance. For comparison, we also indicate the modulus with the highest magnitude: $c_{333} = (-8{,}15 \pm 0{,}18) \cdot 10^{11}\,\mathrm{N/m^2}$. All moduli of elasticity of α-quartz can be found in [Thurston et al 1966] and [Bechmann et al 1969].

3.2.8 Piezoelectric Properties of β-Quartz

The piezoelectric properties of β-quartz are defined by a single independent piezoelectric constant. From its crystal symmetry the following matrix of the piezoelectric coefficients results

$$\begin{pmatrix} 0 & 0 & 0 & d_{14} & 0 & 0 \\ 0 & 0 & 0 & 0 & -d_{14} & 0 \\ 0 & 0 & 0 & 0 & 0 & 0 \end{pmatrix} . \tag{3.14}$$

According to dynamic measurements by Cook and Weissler [Cook 1950], at 612 °C $d_{14} = -1{,}86$ pC/N and the average temperature coefficient in the temperature range 585 ... 626 °C is $TK\,(d_{14}) = -12{,}8 \cdot 10^4/\mathrm{K}$.

The possibility of using β-quartz for piezoelectric resonators was investigated in great detail by White [White 1959]. Some of his findings apply to piezoelectric sensors, too.

Sensor elements made from β-quartz can only be designed for the transverse or the shear effect. The maximum transverse sensitivity is obtained with the orientation XYa45°, resulting in $d'_{12}=-0{,}93$ pC/N. An element of such orientation does not loose its piezoelectric properties near the phase transition from α- to β-quartz.

Until now, no manufacturer seems to have produced a sensor with such an element, although it would offer a quite wide operating temperature range.

3.3 Tourmaline

Tourmaline is, from the chemical point of view, an aluminum-borosilicate. Its relatively complex composition can be described by the formula (Na, Ca) (Mg, $Fe)_3B_3Al_6Si_6$ $(O,OH, F)_{31}$ [Donnay and Barton 1967; Donnay and Buerger 1950]. However, other formulae are also given to describe it [Epprecht 1953; Newnham 1975] because the analysis of tourmaline is quite difficult. Many tourmalines contain Li, Mn, Ti, Fe or Cr, producing characteristic coloration by ion charge transfer. Tourmalines rich in iron are black.

Tourmaline belongs to the ditrigonal-pyramidal class of symmetry 3m and therefore, like α-quartz, to the trigonal crystal system. Its crystals are usually elongated in the direction of the z-axis and are characterized by a pronounced trigonal prism with vertical stripes. The threefold z-axis is a polar axis. Its positive direction is defined in different ways. We denote, following Cady [Cady 1964] as positive end of the z-axis the one at which a negative electric charge appears yielded by a positive deformation (elongation), i.e. for $S_3>0$, in the direction of the z-axis. Mason [Mason 1950, p 213] assigns the positive sign to exactly the opposite direction. Fig. 3.13 represents an idealized tourmaline with Cartesian coordinate system.

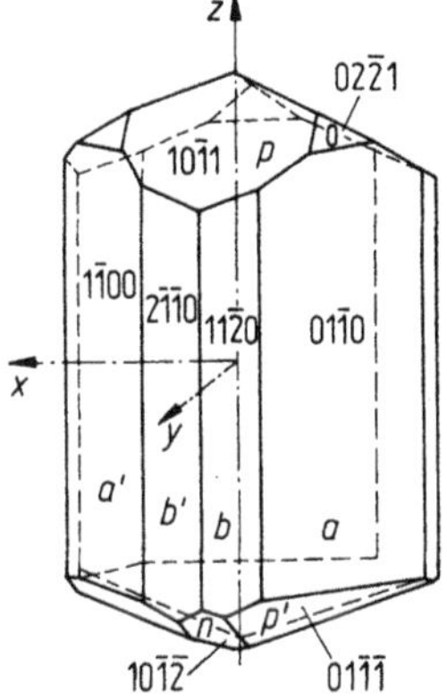

Figure 3.13 Tourmaline crystal [Bechmann and Hearmon 1966]

Table 3.4 Physical properties of tourmaline, lithium niobate and lithium tantalate

Class of symmetry	3m
Tourmaline	$(Na, Ca)(Mg, Fe)_3B_3Al_6Si_6(O, OH, F)_{31}$
Lithium niobate	$LiNbO_3$
Lithium tantalate	$LiTaO_3$

Matrix of elastic coefficients

$$\begin{pmatrix} s_{11} & s_{12} & s_{13} & s_{14} & 0 & 0 \\ s_{12} & s_{11} & s_{13} & -s_{14} & 0 & 0 \\ s_{13} & s_{13} & s_{33} & 0 & 0 & 0 \\ s_{14} & -s_{14} & 0 & s_{44} & 0 & 0 \\ 0 & 0 & 0 & 0 & s_{44} & 2s_{14} \\ 0 & 0 & 0 & 0 & 2s_{44} & 2s_{11}-s_{12} \end{pmatrix}$$

Matrix of elastic moduli

$$\begin{pmatrix} c_{11} & c_{12} & c_{13} & c_{14} & 0 & 0 \\ c_{12} & c_{11} & c_{13} & -c_{14} & 0 & 0 \\ c_{13} & c_{13} & c_{33} & 0 & 0 & 0 \\ c_{14} & -c_{14} & 0 & c_{44} & 0 & 0 \\ 0 & 0 & 0 & 0 & c_{44} & c_{14} \\ 0 & 0 & 0 & 0 & c_{14} & \frac{1}{2}(c_{11}-c_{12}) \end{pmatrix}$$

Matrix of piezoelectric coefficients

$$\begin{pmatrix} 0 & 0 & 0 & 0 & d_{15} & -2d_{22} \\ -2d_{22} & d_{22} & 0 & d_{15} & 0 & 0 \\ d_{31} & d_{31} & d_{33} & 0 & 0 & 0 \end{pmatrix}$$

Matrix of piezoelectric moduli

$$\begin{pmatrix} 0 & 0 & 0 & 0 & e_{15} & -e_{22} \\ -e_{22} & e_{22} & 0 & e_{15} & 0 & 0 \\ e_{31} & e_{31} & e_{33} & 0 & 0 & 0 \end{pmatrix}$$

Matrix of permittivities

$$\begin{pmatrix} \varepsilon_{11} & 0 & 0 \\ 0 & \varepsilon_{11} & 0 \\ 0 & 0 & \varepsilon_{33} \end{pmatrix}$$

Table 3.4 continued

Curie temperature Θ_c in °C	Density ϱ in 10^3 kg m^{-3}
	3,1
1210°C	4,63
665°C	7,454

Elastic coefficients in 10^{-12} $N^{-1}m^2$

	s_{11}^E	s_{12}^E	s_{13}^E	s_{14}^E	s_{33}^E	s_{44}^E	s_{66}^E
Tourmaline	3,85	– 0,48	– 0,71	0,45	6,36	15,4	8,66
$LiNbO_3$	5,78	– 1,01	– 1,47	– 1,02	5,02	17,0	13,6
$LiTaO_3$	4,86	– 0,29	– 1,24	0,63	4,36	10,5	10,3

Elastic moduli in 10^9 Nm^{-2}

	c_{11}^E	c_{12}^E	c_{13}^E	c_{14}^E	c_{33}^E	c_{44}^E	c_{66}^E
Tourmaline	272	40	35	– 6,8	165	65	116
$LiNbO_3$	203	53	75	9	245	60	75
$LiTaO_3$	228	31	74	– 12	271	96	98

Piezoelectric coefficients in 10^{-12} CN^{-1}

	d_{15}	d_{22}	d_{31}	d_{33}	d_h
Tourmaline	3,63	– 0,33	0,34	1,83	2,51
$LiNbO_3$	68	21	– 1	6	4
$LiTaO_3$	26	8,5	– 3,0	9,2	3,2

Piezoelectric moduli in Cm^{-2}

	e_{15}	e_{22}	e_{31}	e_{33}
Tourmaline	0,25	– 0,02	0,10	0,32
$LiNbO_3$	3,7	2,5	0,2	1,3
$LiTaO_3$	2,7	2,0	– 0,1	2,0

Relative permittivities

	$\left(\frac{\varepsilon_{11}}{\varepsilon_0}\right)^T$	$\left(\frac{\varepsilon_{33}}{\varepsilon_0}\right)^T$
Tourmaline	8,2	7,5
$LiNbO_3$	84	30
$LiTaO_3$	53	44

The lattice constants of tourmaline depend on its composition and are at room temperature: $a_0 = 1{,}582 \dots 1{,}599$ nm and $c_0 = 0{,}708 \dots 0{,}720$ nm. The ratio of the lattice constant is $c_0/a_0 = 0{,}447 \dots 0{,}453$ [Epprecht 1953]. The density ϱ varies in the range of $3{,}0 \cdot 10^3$ to $3{,}2 \cdot 10^3$kg/m^3 and increases with the Fe-content.

Tourmaline has a high mechanical strength and, like quartz, is resistant against most acids and alkali.

The most important physical properties of tourmaline are given in Table 3.4. The values were taken from Mason [Mason 1950] and coincide quite well with the much earlier measurements done by Riecke and Voigt [Riecke and Voigt 1892]. The signs of the piezoelectric constants were adapted as defined by Cady [Cady 1964]. Following Mason [Mason 1950], all constants except d_{22} and e_{22} have the opposite sign. No indication about the temperature-dependence of the piezoelectric constants can be found in [Bechmann and Hearmon 1966; Bechmann et al 1969]. Cady [Cady 1964, p 228] mentions that according to Lissauer the value of d_{33} varies less than 2 % in the range −192 ... 19 °C. The small temperature dependence of d_{33} was also confirmed by a summary measurement made during the investigation for his work [Seil 1978]. The piezoelectric coefficient for hydrostatic pressure is $d_h = d_{33} + 2\,d_{31} = 2{,}51$ pC/N and is exploited in sensors for dynamically measuring hydrostatic pressures (e.g. explosion pressures).

Tourmaline offers the advantage that it can not twin and does not have have a phase change below 900 °C, which makes possible to use it in sensors for a very wide temperature range up to about 700 °C (mostly pressure and acceleration sensors). Its biggest disadvantages are the inherent strong pyroelectric effect and the problem of finding crystals of acceptable quality and sufficient quantity from natural occurrences.

According to [Cady 1964] the temperature dependence of the pyroelectric effect p_3 is described by

$$p_3 = \left\{3{,}77 + 0{,}03K^{-1}\left(\Theta - 18°\mathrm{C}\right)\right\}10^{-6}\,\mathrm{Cm}^{-2}\,2K^{-1}. \tag{3.15}$$

Despite the pyroelectric effect, the Z-cut is the most suitable and generally used type of tourmaline element for sensors. The measurement of the electrocaloric effect is reported in [Brüning 1969].

The modulus of elasticity c_{33} in direction of the thickness of a Z-tourmaline plate is about 1,85 times larger than the modulus of elasticity c_{11} of quartz. A further advantage is that the elastic lateral expansion and the thermal longitudinal expansion in the plane normal to the z-axis are isotropic. The Poisson ratio is very small, i.e. $s_{13}/s_{33} = -s_{23}/s_{33} = -0{,}026$. The longitudinal thermal expansion coefficients of tourmaline in the temperature range 0...320°C are defined by the formulae $\alpha_{11} = (3{,}583 + 4{,}490 \cdot 10^{-3}\mathrm{K}^{-1} \cdot \theta) \cdot 10^{-6}\mathrm{K}^{-1}$ and $\alpha_{33} = (8{,}624 + 5{,}624 \cdot 10^{-3}\mathrm{K}^{-1} \cdot \theta) \cdot 10^{-6}\mathrm{K}^{-1}$. Obviously they are much smaller than those of quartz. Natural tourmaline crystals are quite brittle and their properties differ from crystal to crystal. Selecting, working and finishing them in making transduction elements is rather difficult and hence expensive. Attempts to artificially grow tourmaline of usable size have been unsuccessful so far.

3.4 Gallium Orthophosphate

Gallium orthophosphate (gallium phosphate, $GaPO_4$) is a representative of the group of quartz homeotypes. Their crystallographic structure can be derived from that of α-quartz (SiO_2) by substituting the Si atoms alternatingly by a III-valent (such as Al, Fe, Ga) and a V-valent (P, As) element. The unit cell is thus doubled along the z-axis. The point group remains 32, as in α-quartz. However, the space groups are reversed, being $P3_221$ for the optically left-handed forms and $P3_121$ for the optically right-handed forms. Thus, the pyroelectric effect, which is a serious problem especially in high-temperature sensorics, does not exist due to crystallographic symmetry.

The only mineral from this group is berlinite ($AlPO_4$). Because its piezoelectric coefficients are higher than quartz and temperature-compensated bulk and surface wave orientations do exist, its growth and properties were subject of several studies. However, the only quartz homeotype which is presently industrially produced for use in piezoelectric sensors is gallium orthophosphate ($GaPO_4$), which is grown hydrothermally from acid solutions and is being used for pressure sensors for internal combustion engines since 1994.

At room temperature, the lattice constants of $GaPO_4$ are $a = 0{,}4901$ nm and $c = 1{,}1048$ nm and the density is $3{,}57 \cdot 10^3\,kg/m^3$. The α-β phase transition to the point group 622, which is observed in quartz and $AlPO_4$, does not take place. The low-temperature form is stable up to over 950 °C, where a reconstructive transition to a cristobalite-like phase takes place. This has the effect that, contrary to quartz, stress-induced twinning does not occur in technical applications in the whole range of stability.

As in quartz, different conventions for the axes in right- and left-handed $GaPO_4$ can lead to different signs of the piezoelectric and elastic coefficients. The values in Table 3.5 are according to the IEEE Standard 176-1987, which follows the convention introduced by Voigt.

For the applications in question, the most remarkable feature is the temperature stability of the piezoelectric coefficient d_{11}. It does not deviate significantly from its room temperature value of 4,5 pC/N, about twice the value of quartz, up to 500 °C and stays within few percent of that up to 700°C. Above this temperature, no reliable data exist, but a slow decrease until the phase transition temperature is assumed. The temperature dependence of d_{11}, in comparison with that of quartz, is shown in Fig. 3.14. This means that exceptional temperature stability can be achieved for piezoelectric sensors of longitudinal compression type (d_{11}), transverse compression type ($d_{12} = -d_{11}$) and shear type ($d_{26} = -2d_{11}$).

The elastic coefficients were measured in the range of 50 ... 700 °C. Due to the positive temperature coefficients of c_{66} and c_{14}, there are orientations with temperature-compensated frequency for thickness-shear resonators as well as for surface acoustic waves.

The electric resistance in x and y directions and the bending strength are similar to those of quartz.

Table 3.5 Selected piezoelectric properties of $GaPO_4$

General Data		
Point group:	32 (D_3)	
Lattice constants (25°C)	a = 0,4901 nm	c = 1,1048 nm
Density (25°C)	3 570 kg/m^3	
Heat capacity (25°C)	560 J/kgK	
Electric resistivity (25°C)	>100 TΩcm	

Piezoelectric Constants				
pC/N	20°C	500°C	700°C	950°C
d_{11}*	4,5	4,5	4,5	4,1 **
d_{14}	1,9	1,6	1,4 **	1,0 **

Thermal Expansion Coefficients ($T_0 = 25$°C)				
	10^{-6} K^{-1}	10^{-9} K^{-2}	10^{-12} K^{-3}	10^{-15} K^{-4}
α_{11}	12,78	10,6	–16,1	12,3
α_{33}	3,69	5,0	–5,4	3,6

Elastic Constants ($T_0 = 25$°C)				
	GPa	10^{-6} K^{-1}	10^{-9} K^{-2}	10^{-12} K^{-3}
c_{11}^E	66,58	–44,1	–28,5	–59,4
c_{12}^E	21,81	–226,7	–70,8	–205,7
c_{13}^E	24,87	–57,6	41,3	–109,9
c_{14}^{E*}	3,91	507,2	280,6	–99,9
c_{33}^E	102,13	–127,5	–18,3	–134,8
c_{44}^E	37,66	–0,4	–43,8	–37,1
c_{66}^E	22,38	44,9	–7,9	11,9
	$10^{-12}N^{-1}m^2$	10^{-6} K^{-1}	10^{-9} K^{-2}	10^{-12} K^{-3}
s_{11}^E	17,93	400,1	548,8	1122,3
s_{12}^E	–4,82	1013,1	1,4	1311,7
s_{13}^E	–3,19	–577,7	–250,7	–1033,0
s_{14}^{E*}	–2,36	–1138,8	–746,1	–19,0
s_{33}^E	11,35	1673,3	161,9	2977,3
s_{44}^E	27,04	505,5	1479,8	1408,1
s_{66}^E	45,51	–1226,0	1094,8	378,9

Electrooptic Constants (λ = 632,8 nm, 25°C)	
	pm/V
r_{11}^{T*}	–0,95
r_{41}^{T*}	–0,62
r_{11}^{S*}	–0,90
r_{41}^{S*}	–0,05

Relative Dielectric Constants (1 kHz, 25°C)	
ε_{11}^T	6,1
ε_{33}^T	6,6
ε_{11}^S	5,8
ε_{33}^S	6,6

* signs according to standard IEEE 176-1987 for right-handed crystal; for left-handed $GaPO_4$ crystals, the signs of d_{11}, d_{14}, r_{11} and r_{11} are changed. For the use of IEC 758 1993-04 the following values apply for left-handed and right-handed $GaPO_4$ crystals (25°C):

c_{14}	–3,91	d_{11}	–4,5	r_{11}^T	0,95	r_{11}^S	0,90
s_{14}	2,36	d_{14}	1,9	r_{41}^T	–0,62	r_{41}^S	–0,05

** expected values

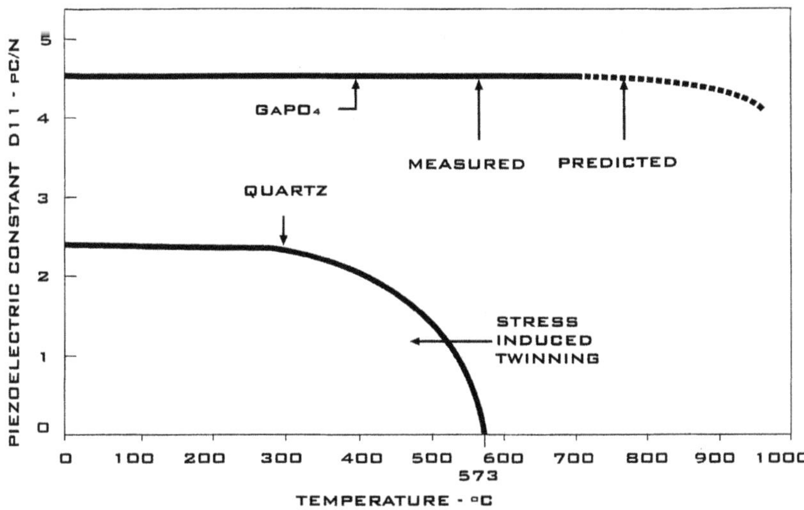

Figure 3.14 Temperature dependence of d_{11} in quartz and gallium orthophosphate (Courtesy of AVL)

More information on this material can be found in [Briot et al 1999; Détaint et al 1997; Krempl, Krispel et al 1997; Krempl, Schleinzer et al 1997; Krispel, Krempl et al 1997; Krispel, Schleinzer et al 1997; Palmier et al 1997; Philippot et al 1993; Reiter et al 2001; Reiter et al 2000; Reiter et al 1997; Thanner et al 2000; Wallnöfer et al 2000; Worsch et al 1998; Worsch et al 2001; Wallnöfer et al 1994]. Additional references and information on gallium orthophosphate can be found on the web site of AVL List GmbH.

3.5 Crystals of the CGG Group

Since the beginning of the 1980's intensive research was done on the synthesis of new piezoelectric crystal compositions. The main objectives consisted in the search for new single crystals with improved piezoelectric properties like high sensitivity, high electro-mechanical coupling, and high stability at elevated temperatures. Among the numerous synthesized crystals, compounds with the $Ca_3Ga_2Ge_4O_{14}$ (CGG) structure were found to possess the most promising properties for piezoelectric applications. Compounds of the CGG structure belong to the trigonal crystal system, point group 32, space group P321 (Schoenflies D_3^2), which is very similar to that of quartz ($P3_121$ and $P3_221$). These crystals are not pyroelectric.

More than 40 single crystal compounds of the CGG group with similar structure were synthesized [Fukuda et al 1998; Mill and Fukuda 1998; Mill et al 1998; Mill et al 1999]. A large number of isomorphous compounds can be obtained by substitution of the cations. At present some piezoelectric data has been reported only for about 10 of these compounds. The full set of elastic, piezoelectric and dielectric constants has been determined for CGG, $La_3Ga_5SiO_{14}$ (LGS), $La_3Ga_{5.5}Ta_{0.5}O_{14}$ (LGT), $La_3Ga_{5.5}Nb_{0.5}O_{14}$ (LGN), $La_3Ga_5GeO_{14}$ (LGG), and $Sr_3Ga_2Ge_4O_{14}$ (SGG). The characterization of these crystals at elevated temperatures is still not complete. In Table 3.6 some piezoelectric properties are presented.

The crystals listed in Table 3.6 are grown from the melt by the Czochralsky method. A description of this method can be found in [Hurle and Cockayne 1993]. Single crystal growth is complicated due to evaporation of some melt components (Ga_2O_3, GeO_2) [Mill et al 1999]. Below the melting point no phase transition was found in these crystals. Therefore, they are good candidates for high temperature sensor applications. Thanks to the high sensitivity in comparison to quartz these crystals are suitable for miniaturization. It has been observed that the piezoelectric coefficient d_{11} of CGG-type crystals shows a tendency to increase with increasing lattice constant a [Fukuda et al 1998]. The high electromechanical coupling coefficients make these crystals also very attractive for frequency control applications.

The most important single crystal of the CGG group is LGS (Langasite) [Mill et al 2000]. Crystal growth of LGS was transferred to industrial production and crystal wafers up to 4" diameter have become commercially available. LGS is one of the most promising new crystals for frequency control applications both for bulk acoustic wave (BAW) and surface acoustic wave (SAW) devices [Grouzinenko and Bezdelkin 1992; Ssakharov et al 1992]. Applications for SAW sensors at high temperatures have been reported [Hornsteiner et al 1998]. The electromechanical properties of LGS were investigated thoroughly with ultrasonic and resonance methods [Bezdelkin and Antonova 1994]. Piezoelectric force measurements up to 400 °C with LGS were performed in [Ganschow et al 1994]. An important property for piezoelectric force, pressure or acceleration sensor applications consists in the behavior of the electric resistance at high temperatures and controlled atmosphere. For optimum operation in sensors the resistance of the crystals has to remain stable and of comparable magnitude to quartz. Results about the high temperature properties of LGS are presented in [Fritze et al 1999].

Among the crystals listed in Table 3.6 especially LGT [Mill et al 1999; Pisarevsky et al 1998] and LGN [Fukuda et al 1998; Pisarevsky et al 1995] have attracted a lot of interest. Its piezoelectric coefficients are higher than for LGS. The feasibility to operate at higher temperatures than quartz opens up new application possibilities. This is especially true for SGG [Kaminskii et al 1984]. This crystal possesses the highest sensitivity (about 4 times that of quartz) of the compounds listed in Table 3.6. However, the crystal growth of SGG is more complicated than for LGS because of the evaporation of GeO_2 [Kochurikhin et al 1997]. The growth of large size crystals is further made more difficult due to common problems with inclusions and crystal cracking. In LGS, LGT and LGN the formation of twins during crystal growth has been observed [Chai et al 1998; Naumov et al 1996]. According to [Mansfeld and Boy 1997], lattice ferrobielastic twinning at very high uniaxial pressure was experimentally found for LGS.

Table 3.6 Selected piezoelectric properties of some single crystal compounds of the CGG group

	Abbre-viation	T °C	d_{11} 10^{-12} CN^{-1}	d_{14} 10^{-12} CN^{-1}	e_{11} 10^{9} NC^{-1}	e_{14} 10^{9} NC^{-1}	References
$Ca_3Ga_2Ge_4O_{14}$	CGG	1370	5,38	–2,58		–0,0014	**a**
$L_3Ga_5SiO_{14}$	LGS	1470	6,16	–5,36	0,45	–0,077	**b, c, d, e**
$La_3Ga_5GeO_{14}$	LGG	1470	6,59	–5,51			
$La_3Ga_{5.5}Ta_{0.5}O_{14}$	LGT	1510			0,54	–0,07	**d**
$La_3Ga_{5.5}Nb_{0.5}O_{14}$	LGN	1510	6,63	–5,55	0,44	–0,05	**c, g**
Sr	SGG	370	9,41	–6,96	0,567	–0,055	**a, e**

T_m: melting point
d_{11}, d_{14}, e_{11}, e_{14}: piezoelectric coefficients (a positive sign was chosen for d_{11} and e_{11} according to the IEEE Standards on Piezoelectricity (ANSI/IEEE Std 176-1987) for right-handed crystals)
References: **a**: Kaminskii et al 1984, **b**: Kochurikhin et al 1997, **c**: Silvestrova et al 1987, **d**: Pisarevsky et al 1998, **e**: Bezdelkin and Antonova 1983, **f**: Silvestrova et al 2000, **g**: Pisarevsky et al 1995

Crystals of the CGG group (Fig. 3.15) group are already used for sensor applications such as force sensors, strain sensors and high-temperature pressure sensors [Gossweiler and Cavalloni 1999]. These sensors use the direct piezoelectric effect and offer increased sensitivity in comparison to quartz sensors, high dynamic ratio and high temperature capability. The single crystal material is grown by the sensor manufacturer himself.

Figure 3.15 Crystal of the CGG Group grown from the melt by the Czochralsky method (Courtesy of Kistler)

Table 3.7 Selected piezoelectric properties of some single crystal compounds

	Abbre-viation	T_m °C	d_{33} 10^{-12} CN^{-1}	d_{14} 10^{-12} CN^{-1}	e_{33} 10^9 NC^{-1}	e_{14} 10^9 NC^{-1}	References
$L_2B_4O_7$	LTB	917	24,0	0	0,9	0	**a, b**
$B_{12}GeO_{20}$	BGO	930	0	37,6	0	0,987	**c**
$B_{12}SiO_{20}$	BSO	895	0	40,0	0	1,0	**d**
ZnO			12,3	0	0,96	0	**e**

References: **a**: Shorrocks et al 1981, **b**: Bhalla et al 1985, **c**: Zelenka 1978, **d**: Schweppe and Quadflieg 1974, **e**: Tokarev et al 1975

Another new single crystal material which is a potential candidate for piezoelectric applications is lithium tetraborate ($Li_2B_4O_7$; LTB) [Bhalla et al 1985; Shorrocks 1981]. This crystal belongs to the tetragonal symmetry 4 mm with a polar axis. The high piezoelectric coefficient d_{33} (see Table 3.7) makes LTB attractive for sensor applications. However, LTB is pyroelectric, but not ferroelectric, and has a relatively low melting temperature of 917 °C. This strongly limits the field of applications for LTB, which will be similar to Lithium niobate or Lithium tantalate. The high electromechanical coupling of LTB makes it attractive for SAW applications.

Other piezoelectric crystals such as $Bi_{12}GeO_{20}$ (BGO) [Ballman 1967; Zelenka 1978] or $Bi_{12}SiO_{20}$ (BSO) [Schweppe and Quadflieg 1974] have not found significant applications in sensors for pressure, force or acceleration. BGO and BSO belong to the cubic space group I23 showing no pyroelectricity and possess one independent piezoelectric coefficient $d_{14}=d_{25}=d_{36}$ (see Table 3.7). Zinc oxide (Zincite, ZnO) has hexagonal symmetry (space group $P6_3mc$) and is mainly grown by hydrothermal synthesis [Tokarev et al 1975] (see Table 3.7). The main application for this material consists in the thin film deposition on various substrate materials (e.g. Silicon) as piezoelectric actuator for resonator structures.

3.6 Other Piezoelectric Single Crystals

Besides quartz, tourmaline, gallium orthophosphate and crystals of the CGG group, there are only few other single crystals which are used in piezoelectric sensors, usually for special applications only. Two of them – lithium-niobate and lithium-tantalate – have the same symmetry as tourmaline and their electromechanical material constants are summarized in Table 3.4. Both ferroelectric single crystals were introduced in the 1960s and have been thoroughly investigated, especially for their application in telecommunications. They combine excellent electromechanical properties with a high piezoelectric sensitivity. The Curie temperature of lithium-niobate is very high (around

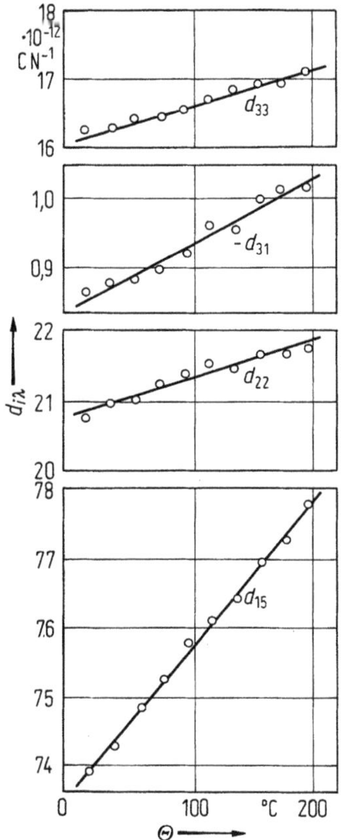

Figure 3.16 Temperature dependence of the piezoelectric coefficients d_{33}, d_{31}, d_{22} and d_{15} in $LiNbO_3$ [Bechmann et al 1969]

1200 °C) which means a relatively low temperature dependence of its material properties (Fig. 3.16). The insulation resistance drops substantially at higher temperature which is a handicap in sensors because it increases the lower frequency limit (see 11.2.3). Therefore sensors with such transduction elements are mainly used in acceleration sensors for high temperature applications.

Although single crystals of Rochelle salt ($NaKC_4H_4O_6 \cdot 4H_2O$) feature a high piezoelectric sensitivity (they were used widely in "crystal pick-ups" for phonographs, until analog records were replaced by the compact disc) they have poor mechanical properties and are highly hygroscopic which is undesirable in sensors. Therefore they are now of historic interest only (e.g. the Russian literature [Sheludew 1975] still refers to ferroelectricity as "Seignette electricity" because Rochelle salt is called "Seignettesalz" in German and "sel de Seignette" in French).

There are only a few other single crystals which may be suitable for use in sensors. Lithium sulfate-monohydrate ($Li_2SO_4 \cdot H_2O$) – often referred to by

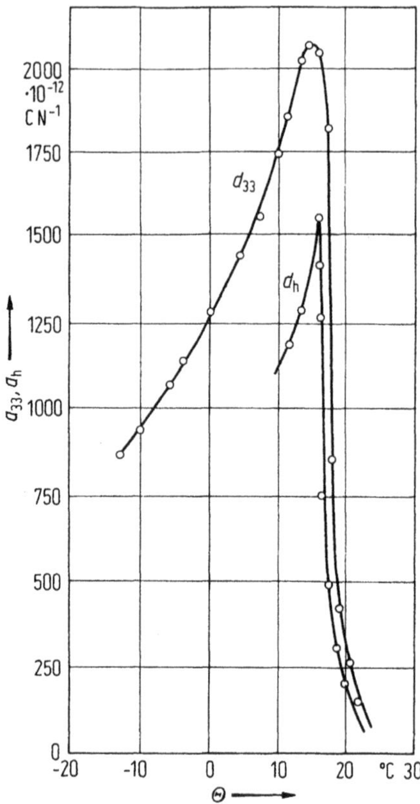

Figure 3.17 Temperature dependence of the piezoelectric coefficients d_{33} and d_h in polycrystalline SbSI [Mitsui et al 1975]

the abbreviation LH or LSH – is such a crystal. Its operating temperature range goes up to only about 90 °C because dehydration starts above 100 °C. LH features relatively high piezoelectric coefficients and a particularly high hydrostatic piezoelectric effect (d_h = 16,4 pC/N) which makes it suitable for pressure sensors for measuring dynamic hydrostatic pressure. A still higher hydrostatic piezoelectric effect is exhibited by the ferroelectric semiconductor SbSI yet its Curie temperature is only 22 °C, too low for most practical applications (Fig. 3.17).

More exhaustive information on the material constants of piezoelectric crystals, including the ones mentioned here, can be found in [Bechmann and Hearmon 1966; Bechmann et al 1969; Berlincourt 1971; Berlincourt et al 1964; Mitsui et al 1975; Mitsui et al 1969]. For many crystals, the temperature dependence of different material constants is given, too. The values given here have been taken from these sources. [Bechmann and Hearmon 1966; Bechmann et al 1969] whose revision and updating is in process [Landolt-Börnstein 1979].

3.7 Piezoelectric Textures

Textures are macroscopically homogenous media composed of a large number of particles, arranged and oriented spatially in a regular pattern. Although they exhibit a number of physical properties characteristic of crystals, they do not have a macroscopic crystal structure. Examples for textures are crystalline textures, consisting of oriented crystals, fiber materials (e.g. wood), electrets which contain oriented dipoles, and piezoelectric ceramics with a preferred direction of spontaneous polarization in the domains of single crystals. In general, textures can be isotropic or anisotropic.

Shubnikow [Shubnikov 1946; Shubnikov et al 1956] studied the piezoelectric properties of textures in great detail, especially from the point of view of their symmetry. The symmetry of textures is determined by the symmetry of the building blocks (particles) and by the symmetry of their mutual arrangement. Textures with the least symmetry are obtained through a parallel arrangement of the crystallites with their triclinic symmetry. Its anisotropy corresponds to that of the triclinic crystal system. However, if the same crystallites are arranged in such a way that they are oriented in parallel with a certain single crystal direction (axis) among each other only, but rotated about this direction (axis) through an arbitrary angle, the texture has a symmetry axis of an infinitely high order and belongs to the group ∞. Out of the 7 infinite symmetry groups only 3 have no symmetry center and therefore can be piezoelectric. These are: ∞. ∞ mm and ∞ 2.

3.7.1 Piezoelectric Ceramics

Piezoelectric ceramics are of particular importance for piezoelectric sensors. They are obtained by sintering a finely ground powdered mixture preferably made of ferroelectrics of the oxygen-octahedral type which first has been shaped into the required form by compressing. Piezoelectric ceramics consist of a large number of ferroelectric grains (crystalllites), each containng domains in which the electric dipoles are aligned. The single domians are at first randomly oriented and mostly have the symmetries mm2, 3m or 4mm, yet their totality forms a texture belonging to the symmetry group $\infty / \infty /$ mmm. It is isotropic and has no piezoelectric properties.

In order to show piezoelectric properties, the ceramic must be polarized (or *poled*). This is achieved at a high temperature (usually around 200 °C) by applying a strong electric field through which the spontaneous polarization of the individual micro crystals is partially aligned with the direction of the field. After polarization the polar axes of the various single crystals are oriented within a certain solid angle whose magnitude is determined by the mechanism of polarization. Through a complex interaction between the aligned domains in the crystallites and the alien phases contained in the structure, the direction of the resulting spontaneous polarization and the corresponding piezoelectric properties of the ceramic as well persist after the electric field is switched off. A polarized

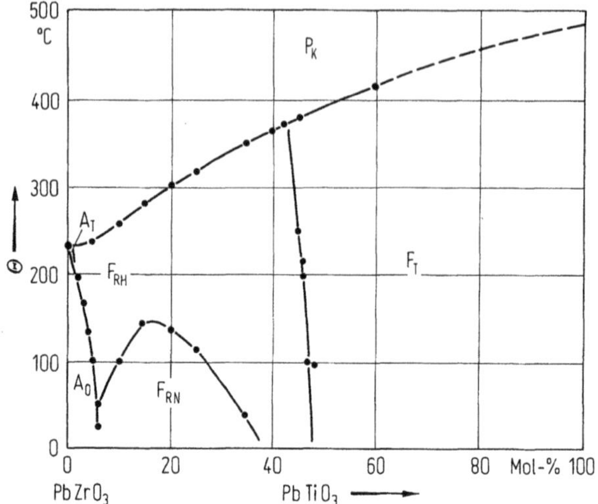

Figure 3.18 Phase diagram of the solid solution $PbZrO_3$–$PbTiO_3$. P_K paraelectric-cubic phase, F_{RH} ferroelectric-rhombohedral high-temperature phase, F_{RN} ferroelectric-rhombohedral low-temperature phase, F_T ferroelectric-tetragonal phase, A_o antiferroelectric-rhombic (pseudo-tetragonal) phase, A_T antiferroelectric-tetragonal phase.

ceramic plate has macroscopic properties similar to a crystal with a ∞-counting axis of rotation. Therefore it is possible to polarize ceramic rings, cylinders, hollow cone and sphere sections, and thus to produce piezoelectric elements of almost any shape and size [Bauer et al 1976; IRE 1961; Jaffe et al 1971].

In the late 1940s the first experiments with piezoelectric barium-titanate ceramics were made. Their main shortcoming is their low Curie temperature of only about 120°C, severely limiting their operating temperature range.

The most widely used ferroelectric ceramics are now those consisting of solid solutions of $PbZrO_3$ and $PbTiO_3$ [Berlincourt 1971; Berlincourt et al 1964; Carl 1972; Feldtkeller 1974; Randeraat and Setterington 1974]. They are designated as lead-zirconite-titanate mixed ceramics (PZT). The phase diagram of this system is shown in Fig. 3.18. At high temperatures all materials of this series of mixtures have the cubic perovskite lattice like $BaTiO_3$. Below the Curie temperature which depends on the $PbTiO_3$-content mixtures rich in $PbTiO_3$ take a tetragonal crystal

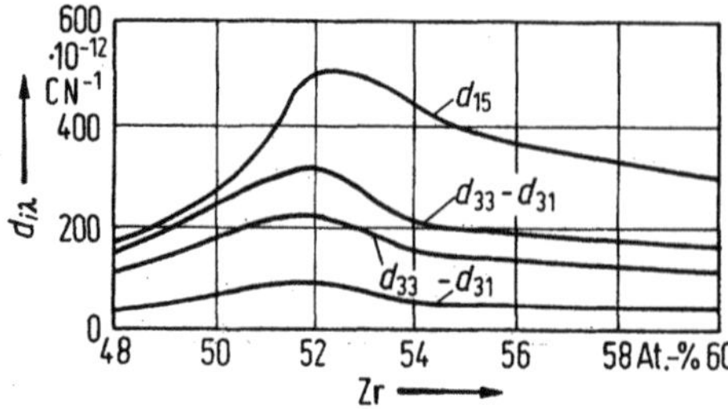

Figure 3.19 Variation in the piezoelectric coefficients of the mixed ceramics $Pb(Zr,Ti)O_3$ with a Zircon content from 48 to 60% of atoms [Bauer et al 1976]

Table 3.8 Selected piezoelectric properties of commercially available and new high temperature piezoelectric ceramics (references for these materials can be found in reference [N.N. 1994])

	T_{Curie} °C	d_{33} pC/N	d_{31} pC/N	d_{15} pC/N	g_{33} 10^{-3}Vm/N	g_{31} 10^{-3}Vm/NN	g_{15} 10^{-3}Vm/N	k_{33}	Ref
PZT family									
PZT (Hard)	330	190	–55		54	–16		0,67	**a**
PZT (Soft)	250	425	–170	500	27	–11		0,70	**a**
Lead metaniobate									
$PbNb_2O_6$	500	100			43				**a**
$(BaPb)Nb_2O_6$	400	85		100	32		46	0,30	**c**
Bismuth titanate family									
Bismuth titanate	650	18	–2	16	17	–2		0,09	**a**
Bi_3TiNbO_9	940	5							**b**
$SrBi_2Nb_2O_9$	440	10							**b**
$Na_{0,5}Bi_{4,5}Ti_4O_{15}$	655	10							**b**
$SrBi_4Ti_4O_{15}$	530	15							**b**
$CaBi_4Ti_4O_{15}$	787	14							**b**
Perovskite layer structure									
$Sr_2(Nb_{0,5}Ta_{0,5})_2O_7$	820			2,6					**c**
Thin film									
AlN		5,5	–2,65	–4,1	66 ... 105				**d, e**
ZnO		10,6			135				**d**

References: **a:** N.N. 2001, **b:** Voisard 2000, **c:** N.N. 1994, **d:** Patel and Nicholson 1990, **e:** Tsubouchi 1981

structure, while mixtures poor in $PbTiO_3$ go into a rhombohedral phase. In compositions between $52PbZrO_3/48PbTiO_3$ and $56PbZrO_3/44PbTiO_3$ both ferroelectric phases can coexist. In this range, with a suitable composition of the ceramic, many material properties show extreme values.

Mixed ceramics of this composition can be readily polarized and then exhibit high piezoelectric coefficients (Fig. 3.19 and Table 3.8). The highest possible piezoelectric sensitivity is achieved with a solid solution with 55 % lead zirconate and 45 % lead titanate, sometimes doped with lanthanum. The piezoelectric sensitivity that can be achieved is so high that the direct piezoelectric effect can be used to build igniters for lighters and for heating systems. The extremely high charge yield obtained from a mechanical strike combined with the inherent low capacitance of such elements result in such high voltages that a spark is easily produced. Yet attempts to use such elements for fuel ignition in internal combustion engines have failed so far, mainly because of the limited life and stability of such elements under repeated mechanical striking.

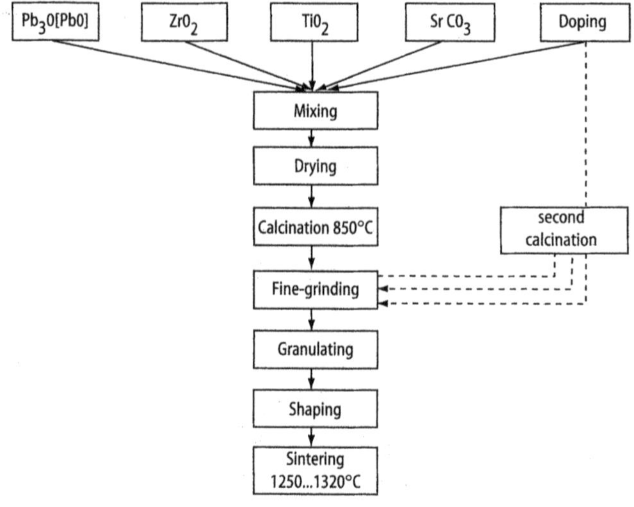

Figure 3.20 Example of a processing sequence for making lead-circonate-titanite-mixed ceramics [Bauer et al 1976]

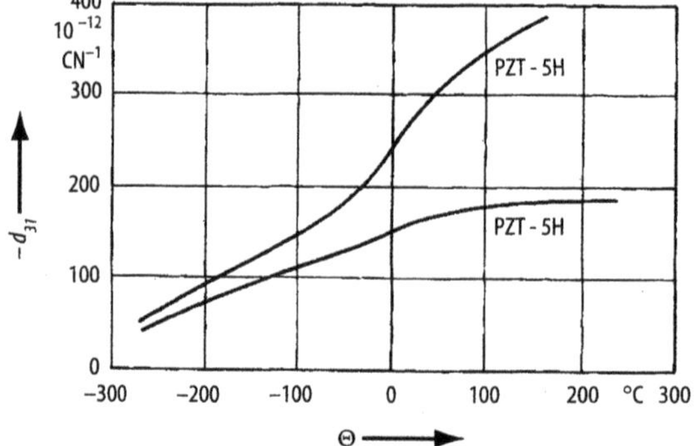

Figure 3.21 Temperature dependence of the piezoelectric coefficient d_{31} in PZT 5A and PZT 5H [Bechmann and Hearmon 1966]

Fig. 3.20 shows schematically the process for making lead-zirconate-titanate-mixed ceramics.

A drawback of all ceramics is that they suffer from stronger aftereffects and fatigue than single crystals. Also their material properties are strongly temperature-dependent (Fig. 3.21).

Advantages of ceramics over single crystals are:

- *Low cost*: producing and processing ceramics is much cheaper and faster than growing, cutting and machining single crystals. Ceramics can easily be produced in large quantities and in an almost unlimited variety of shapes, requiring only minimal machining.
- *Availability*: some single crystals are very difficult to grow artificially or can not be grown at all.
- *Piezoelectric sensitivity*: some piezoelectric ceramics exhibit piezoelectric coefficients up to over a hundred times higher than those observed in single crystals.

Drawbacks of piezoelectric ceramics are:

- *Stability*: the polarizing (also called *poling*) process can be difficult (even leading to micro cracks in the ceramic material). Under load at higher temperature, the degree of polarization may slowly decrease, resulting in a slow decrease in sensitivity over time. This phenomenon is also called *aging* and can be minimized by first artificially aging the ceramic elements (usually subjecting the completed sensor to a number of temperature and load cycles before final calibration).
- *Temperature coefficient*: in ferroelectric materials, the electromechanical properties usually strongly depend on temperature, especially near the Curie temperature.
- *Temperature range*: the useful temperature range is limited by the Curie temperature, above which the ceramic looses its polarization and therefore all piezoelectric properties.
- *Pyroelectricity*: all ferroelectric materials are pyroelectric, which means that they are also sensitive to temperature changes. The pyroelectric signal can not be distinguished from the piezoelectric signal. By exploiting the piezoelctric shear effect in some materials, the influence of the pyroelectric effect can be largel eliminated.
- *Resistivity*: resistivity (electrical insulation) of ceramics is generally lower than that of single crystals, which means that ceramic sensors can not be used for quasistatic measurements.

PZT is the most common commercially available piezoelectric ceramic for actuator and sensor applications. It exhibits very high piezoelectric coupling coefficients and can be used up to about 200 °C. For high temperature applications (above 200°C), there are several families of material from which a suitable ceramic may be chosen (Table 3.9):

Bismuth titanate family: also known as bismuth layer structure ferroelectrics, enhanced piezoelectric properties may be obtained by modification of one or more of a large number of elements.

Tungsten bronze family: Lead metaniobate ($PbNb_2O_6$) is essentially used in transducers for non-destructive testing and medical diagnostic imaging. It has a very low Q_m and quite high piezoelectric coefficients.

Perovskite layer structure ferroelectrics: similar to the bismuth titanate family, these materials also possess a layered structure. The two best known compounds $Sr_2Nb_2O_7$ and $La_2Ti_2O_7$ exhibit the highest known Curie temperatures (1342°C and 1500°C respectively). The solid solution $Sr_2(Nb_{0,5}Ta_{0,5})_2O_7$ has been shown to possess the highest resistivity of any ceramic materials tested [N.N. 1994].

AlN thin films: AlN is a non-ferroelectric material, which exhibit piezoelectricity activity as thin film, but not as bulk material. Used in a SAW (surface acoustic wave) device, AlN has been shown to exhibit piezoelectric activity at temperature up to 1150°C [Patel and Nicholson 1990].

For special applications at high temperatures piezoelectric ceramics with a high Curie temperature were developed. Especially favorable properties are found in

Table 3.9 Physical properties of mixed ceramics

Class of symmetry 6mm and ∞ m Polarized piezoelectrics ceramics PZT–5A PZT–5H

Matrix of elastic coefficients

$$\begin{pmatrix} s_{11} & s_{12} & s_{13} & 0 & 0 & 0 \\ s_{12} & s_{11} & s_{13} & 0 & 0 & 0 \\ s_{13} & s_{13} & s_{33} & 0 & 0 & 0 \\ 0 & 0 & 0 & s_{44} & 0 & 0 \\ 0 & 0 & 0 & 0 & s_{44} & 0 \\ 0 & 0 & 0 & 0 & 0 & 2(s_{11}-s_{12}) \end{pmatrix}$$

Matrix of ėlastic moduli

$$\begin{pmatrix} c_{11} & c_{12} & c_{13} & 0 & 0 & 0 \\ c_{12} & c_{11} & c_{13} & 0 & 0 & 0 \\ c_{13} & c_{13} & c_{33} & 0 & 0 & 0 \\ 0 & 0 & 0 & c_{44} & 0 & 0 \\ 0 & 0 & 0 & 0 & c_{44} & 0 \\ 0 & 0 & 0 & 0 & 0 & \frac{1}{2}(c_{11}-c_{12}) \end{pmatrix}$$

Matrix of piezoelectrics coefficients

$$\begin{pmatrix} 0 & 0 & 0 & 0 & d_{15} & 0 \\ 0 & 0 & 0 & d_{15} & 0 & 0 \\ d_{31} & d_{31} & d_{33} & 0 & 0 & 0 \end{pmatrix}$$

Matrix of piezoelectrics moduli

$$\begin{pmatrix} 0 & 0 & 0 & 0 & e_{15} & 0 \\ 0 & 0 & 0 & e_{15} & 0 & 0 \\ e_{31} & e_{31} & e_{33} & 0 & 0 & 0 \end{pmatrix}$$

Matrix of permittivities

$$\begin{pmatrix} \varepsilon_{11} & 0 & 0 \\ 0 & \varepsilon_{11} & 0 \\ 0 & 0 & \varepsilon_{33} \end{pmatrix}$$

Table 3.9 continued

Curie temperature Θ_c in °C	Density ϱ in 10^3 kg m^{-3}
365 °C	7,75
193 °C	7,5

Elastic coefficients in 10^{-12} N^{-1}m^2

	s_{11}^E	s_{12}^E	s_{13}^E	s_{33}^E	s_{44}^E	s_{66}^E
PZT – 5A	16,4	– 5,74	– 7,22	18,8	47,5	44,3
PZT – 5H	16,5	– 4,78	– 8,45	20,7	43,5	42,6

Elastic moduli in 10^9 Nm^{-2}

	c_{11}^E	c_{12}^E	c_{13}^E	c_{33}^E	c_{44}^E	c_{66}^E
PZT – 5A	121	75,4	75,2	111	21,1	22,8
PZT – 5H	126	79,5	84,1	117	23,0	23,2

Piezoelectric coefficients in 10^{-12} CN^{-1}

	d_{15}	d_{31}	d_{33}	d_h
PZT – 5A	584	–171	374	32
PZT – 5H	741	–274	593	45

Piezoelectric moduli in Cm^{-2}

	e_{15}	e_{31}	e_{33}
PZT – 5A	12,3	–5,4	15,8
PZT – 5H	17	–6,5	23,3

Relative permittives

	$\left(\frac{\varepsilon_{11}}{\varepsilon_0}\right)^T$	$\left(\frac{\varepsilon_{33}}{\varepsilon_0}\right)^T$
PZT – 5A	1730	1700
PZT – 5H	3130	3400

the compound $PbTiO_3$+ 1% MnO_2 which has a Curie temperature of 520 °C [Bauer et al 1976]. However, the use of this ceramic in piezoelectric sensors is limited by the sharp drop of its insulation resistance with increasing temperature, resulting in a rapid rise of the lower frequency limit.

3.7.2 Piezoelectricity in Thin Films

Although the first observations by Brain were made over 70 years ago, the study of the piezoelectric effect in polymers and biological materials really started only in the late 1960s [Kawai 1969]. In oriented thin films piezoelectric coefficients superior to the piezoelectric sensitivity of α-quartz can be achieved. The practical applications are hampered by the limited stability of the material properties and by relaxation effects. Nevertheless newer investigations of piezoelectric properties of thin film polymers point to interesting potential applications [Hayakawa 1973; Erhart et al 2002]. PVDF (polyvinylidene fluoride) films are used e.g. as sensors in fluid dynamics [Galassi et al 2000; Nitsche et al 1987; Nitsche et al 1988].

Piezoelectric thin films can be used for MEMS (micro electro-mechanical systems). Deposition and patterning of piezoelectric thin films on a substrate (mostly silicon) is achieved with common microstructuring processes, suitable for mass-production. Among the many piezoelectric materials, mainly ZnO, PZT [Polla and Francis 1996] and AlN are potential candidates for sensor applications. ZnO and AlN have to be deposited as single crystalline films due to their non-ferroelectric behavior. On the other hand, PZT can be deposited as polycrystalline film with subsequent poling process.

4 Piezoelectric Sensor Terminology

The field of measuring – like all types of technical activities – calls for well and clearly defined terms. Only if specifications are given in an universally understandable way, leaving no ambiguities whatsoever about their meanings and implications can misunderstandings and misinterpretations be avoided.

Until now there exists unfortunately no unified terminology in the field of the measuring technique for mechanical measurands. One of the most complete collections of terms and definitions for electric sensors is the Standard ANSI/ISA-S37.1-1975 (R1982) entitled "Electrical Transducer Nomenclature and Terminology". It was drawn up by the ISA (Instrument Society of America) during the late sixties when the field of electric measuring of mechanical measurands was developing rapidly, mainly within the field of aeronautics and space exploration [ISA 1982; Norton 1969]. Although somewhat dated (it was reaffirmed in 1982), this standard is still valid and of practical value.

There is no equivalent standard in the VDI/VDE publications. The guide VDI/VDE 2600 (sheets 1-6): "Metrologie (Meßtechnik)", and DIN 1319 (sheets 1–4): "Grundlagen der Meßtechnik (Fundamentals of metrology)" refer only to general terms in the field of measurement. Only e.g. DIN 16086:1992 05 "Elektrische Druckmeßgeräte, Druckaufnehmer, Druckumformer, Druckmeßgeräte; Begriffe, Angaben in Datenblättern (Electrical pressure measuring devices, pressure sensors, pressure measuring converters, pressure measuring devices; terms, specifications in data sheets)" covers in more detail pressure sensors only. However it refers mainly to sensors based on transduction principles other than piezoelectric.

Various sensor manufacturers have their own terminology which sometimes differ greatly from each other or are aimed at a particular type of sensors only.

The ANSI/ISA-S37.1 standard uses the term *transducer* and prefers it over the term *sensor*. However, the technical language has evolved over the 25 years since that standard was written and the term *sensor* is now much more widely used. Therefore, the term *sensor* is consistently used in this book (including in the definitions cited from the standard where it was substituted for *transducer*). In this chapter only those terms and definitions are listed which are relevant to the field of piezoelectric sensors. For more details please refer to the complete standard [ISA 1975] and to [ISA 1982; Norton 1969 and 1982].

A number of other expressions are still widely used to designate a sensor, such as *meter, pickup, detector, cell, converter* and so on. However their use is discouraged because they are not accurate or often even wrong. Some of these terms are simply everyday language (speedometer, tachometer) or even technical slang (pressure pickup, accels, etc.). Accelerometer, flowmeter, tachometer and so on are widely used. "-meter" may be appropriate for e.g. speedometer, barometer or thermometer, but should be avoided in technical fields. Pickup is rather technical slang and should not be used instead of sensor. Detector is a device which just signals the presence but not the magnitude of a measurand. Cell is often used for force sensors of a specific design (load cells) but here too, the preferred term is force sensor. Transmitter is not correct unless it applies to the particular type of sensor with built-in electronics for transmitting the output signal by 2- or 3-wire circuits.

Clear and precise terminology is important to avoid misunderstandings in the communication between manufacturer and user, especially in the international sensor market. The same applies to specifications, i.e. sensor characteristics and their definition. Specific examples are given under the various terms.

For the correct use of names and symbols for quantities and units, see [ISO 1993]. The basic vocabulary for legal metrology and traceability is set down in [ISO et al 1993].

4.1 Definition of a Sensor

Measurand:	A physical quantity, property or condition which is measured.
Sensor:	A device which provides a usable output in response to a specified measurand.
Output:	The electrical quantity, produced by a sensor, which is a function of the applied measurand.
Sensing element:	That part of a sensor which responds directly to the measurand.
Transduction element:	The electrical portion of a sensor in which the output originates.

These five terms define the nature of a sensor. A sensor establishes an unambiguous relation between a measurand acting on it and its output. Usually, a linear dependence of the output on the measurand is aimed at, however there are a number of sensor types having a nonlinear characteristic.

The measurand acts on the sensing element and produces the output in the transduction element. The sensing element is not only that part of a sensor which transmits the action of the measurand but that part whose response allows the measurand in principle to be measured.

In most types of sensors it is easy to distinguish the sensing element from the transduction element (e.g. a flexing beam as the sensing element and the attached strain gage as the transduction element).

Sensors can be divided into 2 groups, called *active* sensors and *passive* sensors. Active sensors yield an output without the need for external energy. Piezoelectric sensors are of the active type while e.g. resistive sensors (with strain gage or piezoresistive transduction elements) are passive. The output of a piezoelectric sensor is a change in polarization which yields an electric charge as output that can be measured directly without external energy. The output of a *passive* sensor is e.g. a change in resistance, capacitance or inductance which requires application of external energy (electric voltage or current) in order to measure it. The output then has the form of a voltage or a current which is a function of the measurand.

Piezoelectric sensors are mainly used for the measurands *force, strain, pressure, acceleration and acoustic emission.* These and other measurands such as temperature, humidity, radiation, displacement, flow rate and so on can also be measured with sensors based on other transduction principles which are described e.g. in [Doebelin 1989; Gohlke 1955; Grave 1962; Neubert 1975; Norton 1969 and 1982; Pflier et al 1978; Profos and Pfeifer 1994 and 1997; Roberts 1946; Rohrbach 1967; Selbstein 1950].

4.2 Properties of Sensors

4.2.1 Static Characteristics

Static properties are those which are not a function of time.

4.2.1.1 *Properties Relating to the Measurand*

Range: The measurand values, over which a sensor is intended to measure, specified by their upper and lower limits.

Span: The algebraic difference between the limits of the range.

Overload: The maximum magnitude of measurand that can be applied to a sensor without causing a change in performance beyond specified tolerance.

Burst pressure: The pressure which may be applied to the sensing element or the case (as specified) of a sensor without rupture of either the sensing element or sensor case as specified.

Fig. 4.1 illustrates the relation between these properties.

The range of a sensor can be unipolar (only positive or negative values of the measurand) or bipolar. Examples for unipolar ranges are pressure sensors (e.g.

0 … 200 bar) or force sensors for compressive forces only (e.g. 0 … 50 kN). Bipolar ranges may be symmetric or asymmetric. Acceleration sensors usually have symmetrical ranges (e.g. –50 … 50 *g*, which often is written as ±50 *g*), except some sensors designed for measuring high shocks (e.g. –2 000 … 100 000 *g*). Within its range the sensor measures within the specified tolerances.

Overload is an important characteristic because it defines the limit(s) within which the proprieties of the sensor will not be changed permanently beyond the specified tolerances valid within the range. Once overload has been exceeded, the sensor will no longer perform within the specified tolerances. As long as overload is not exceeded by the applied measurand, the sensor will measure again within the specified tolerances when the measurand is again inside the range. Overload exceeded or not is an essential criteria for the sensor manufacturer for accepting a warranty claim.

Overload should preferably be specified as the value of the measurand representing the overload (e.g. Range: 0 … 100 bar, overload: 140 bar; Fig. 4.1). All too often overload is specified in percent of the span which always bears the risk of being misinterpreted. In the example just given, the correct value (according to the definition) for the overload must be specified as 140% (140 bar = 140% of 100 bar, the span). Many people would consider this to mean than the range can be exceed by another 140% which would result in 240 bar, clearly outside the specified overload of 140 bar and most likely damaging or even destroying the sensor. Percentage values are especially misleading with bipolar ranges because it may not be clear if the percentage applies to the span (as per definition) or to only the positive or negative part of the range. Example: An acceleration sensor, having a

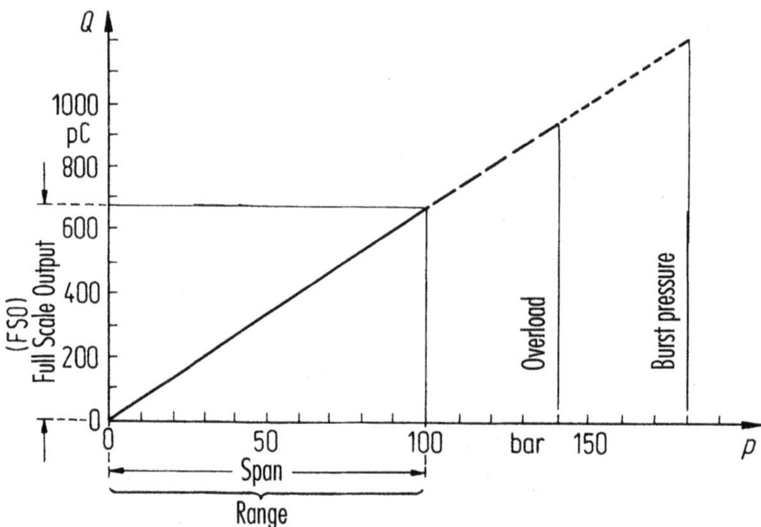

Figure 4.1 Range, overload and burst pressure (exemplified with a pressure sensor). Range: 0 … 100 bar. Overload: 140 bar. Burst pressure: 180 bar. Lower end point: 0 pC. Upper end point: 675 pC. Full scale output (FSO): 675 pC. Span: 675 pC

range of 100 ... 100g (±100g) which means a span of 200g, has a specified overload of 150/150g. Correctly (as per definition) indicated in percent, overload is 75/75% (150g is 75% of 200g, the span!). If a user would read this to mean that the overload is 100g plus 75%, then he would apply 175g, clearly too much. This example demonstrates that overload indications in percent are ambiguous in practice and that it is much better to specify the measurand values valid for overload directly.

Sensors should be selected such that overload never is reached during regular operation. Often exceptional operating conditions must be considered where overload is accepted to be exceeded (with subsequent replacement of the sensor). However safety aspects may dictate that the sensor must not burst because e.g. toxic or aggressive fluids could leak out. Therefore burst pressure is often specified for pressure sensors or other types of sensors mounted with their case exposed to fluid pressure, too.

4.2.1.2
Properties of the Relationship Between Measurand and Output

Calibration:	A test during which known values of measurand are applied to the sensor and corresponding output readings are recorded under specified conditions.
Calibration cycle:	The application of known values of measurand and recording of the corresponding output readings, over the full (or specified portion of the) range of a sensor in an ascending and descending direction.
Calibration curve:	A graphical representation of the calibration record.
Calibration record:	A record (e.g., table or graph) of the measured relationship of the sensor output to the applied measurand over the sensor range. Note: A calibration record which is issued in compliance with internationally recognized standards is called *Calibration Certificate.*
Calibration certificate:	A calibration record, issued by an accredited calibration facility and certifying its traceability to national and international standards (note: this term and its definition is not yet part of the ANSI/ISA-S37.1 standard).
Calibration traceability:	The relation of a *Transducer Calibration*, through a specified step-by-step process, to an instrument or group of instruments calibrated by the National Bureau of standards. Note: The estimated *Error* incurred in each step must be known.

	Remark: This definition will have to be revised with the advent of the introduction of new international standards and regulations on traceability and mutually recognized calibration certificates within the scope of the ISO/IEC/EN 17025 standard [ISO/IEC 1999] on calibration of measuring instruments.
Threshold:	The smallest change in the measurand that will result in a measurable change in sensor output. Note: the term *Resolution* is defined as "The magnitude of output step changes as the measurand is continuously varied over the range". It refers mainly to potentiometric sensors and sensors with digital output (i.e. an output varying in discrete steps) and it is not a correct substitute for the term threshold (see also *Sensitivity*).
Sensitivity:	The ratio of change in sensor output to a change in the value of the measurand. Note: In the sense of the smallest detectable change in measurand use threshold.
Hysteresis:	The maximum difference in Output, at any measurand value within the specified range, when the value is approached first with increasing and then with decreasing measurand. Note: Hysteresis is expressed in percent of full scale output, during any one calibration cycle.
Linearity:	The closeness of a calibration curve to a specified straight line. Note 1: Linearity is expressed as the maximum deviation of any calibration point from a specified straight line, during any one calibration cycle. It is expressed as "within ±... %FSO (Full Scale Output)". Note 2: Terms such as "non-linearity" or "linearity deviation" should no longer be used.

The relationship between measurand and output is determined by calibration. For calibrating, a reference sensor or a device such as a force balance, shaker, pressure generator is required which generates the desired measurand values within known error limits or – as it is more correct to say – with known measuring uncertainty. Such references for measurands must be traceable to international standards.

With the advent of the ISO 9000:2000 series quality management systems, traceable calibration of measuring instruments – which includes sensors – has become a sine qua non. Complete information can be obtained through the *International Laboratory Accreditation Cooperation* ILAC (http://www.ilac.org). On their website, exhaustive information on virtually all organizations worldwide (the "national bureaux of standards") concerned with testing, calibration, traceability and accreditation can be found.

During a calibration cycle measurand values are applied, starting at the lower end point of the range and increasing in magnitude up to the upper end point and back again, and the corresponding output is recorded. The measurand can be varied continuously or applied in discrete steps. Continuous calibration is usually done with a reference sensor while step calibration is done with devices such as dead weight testers, force balances and so on. The result is often visualized by the calibration curve, which allows to determine sensitivity, linearity and hysteresis (Fig. 4.2). Since calibration values are now mostly recorded by data acquisition systems, computing algorithms for determining sensitivity, linearity and hysteresis have been developed.

The threshold describes the smallest change in measurand that can be measured with a given sensor. With most types of sensors and certainly with piezoelectric sensors, the threshold is determined by the amplifier used. In general, a signal corresponding to 3 times the value of amplifier noise is specified as threshold of the sensor. Often the term resolution is used incorrectly to mean the threshold. Resolution describes a step-wise output of a sensor in response to a continuously varying measurand. This is observed only in e.g. potentiometric sensors with wire-wound resistors) and sensors with digital output [ISA 1982; Norton 1969].

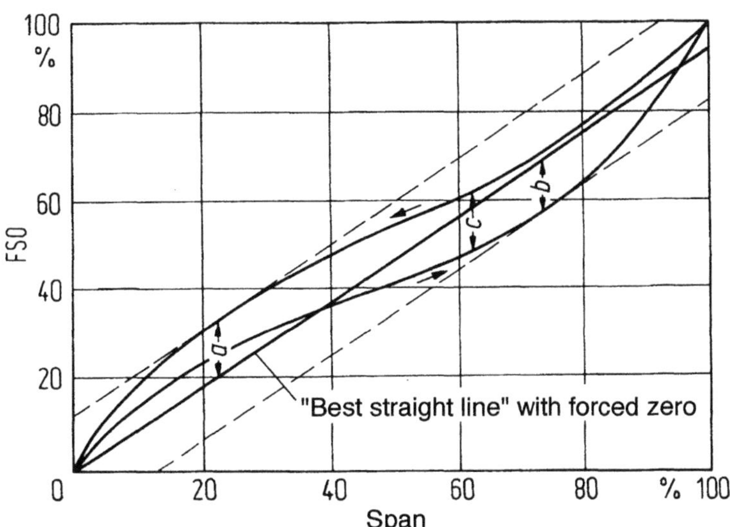

Figure 4.2 Calibration curve, linearity and hysteresis. As an example, independent linearity with forced zero is shown. The deviation of the calibration curve from the reference straight line in the direction of the ordinates has been exaggerated to make the drawing more clear. Linearity: $+a/-b$%FSO. By definition, for independent linearity $a=b$ and therefore the linearity is then $\leq \pm a$%FSO. Hysteresis is $\leq c$%FSO

The output of piezoelectric sensors is always electric charge. Sensitivity is expressed as the ratio of electric charge to measurand. The charge yielded per mechanical unit (N, bar, *g*, με, etc.) is of the order of pC (Picocoulomb) which is the reason why the "pC" has become the generally accepted working unit for electric charge in the field of piezoelectric sensors.

In the strict sense of the term, *sensitivity* is the slope of the tangent on the calibration curve in a given point. If the calibration curve is not a straight line, this means that the sensitivity varies as a function of the measurand value. Although modern electronics and software allow for compensation of such non-linear characteristics, a linear relationship, i.e. a constant sensitivity over the entire range, is still preferred. Most transduction elements and especially the piezoelectric ones have an almost perfect linear characteristic and therefore the sensitivity is generally defined as the slope of the straight line that serves to define and calculate the linearity. Piezoelectric sensors and in particular those with transduction elements made of quartz have a very good linearity and also an exceptionally high ratio of span-to-threshold (often termed dynamic ratio) which can reach values of over 10^8 (e.g. quartz force sensors with a range of MN that still can measure mN, too!). Such sensors are often calibrated not only over their range, but over partial ranges (e.g. 10% and 1%) too, and the corresponding sensitivities are determined. Throughout the book, sensitivity is always understood as the slope of the straight line serving to define the linearity, unless indicated otherwise.

An ideal sensor would yield a definite output for any measurand value, irrespective of whether that value has been reached by an increasing or a decreasing measurand. This would mean that the ascending and the descending part of the calibration curve are congruent. In reality that is rarely so because of hysteresis inherent of certain piezoelectric materials (such as piezoelectric ceramics) or originating from the mechanical construction of the sensor. Usually the specified hysteresis is the largest difference observed at any measurand value within the range and it is expressed in percent of FSO (Full-Scale Output).

A linear relationship between output and measurand, i.e. the output is proportional to the measurand, would give a calibration curve that is a straight line. The term *linearity* describes how much the calibration curve deviates from a certain straight line. Unless it is clearly specified how that straight line is defined, the simple term linearity is meaningless (Fig. 4.3).

Independent linearity:	Linearity referred to the "best straight line".
Best straight line:	A line midway between the two parallel straight lines closest together and enclosing all output versus measurand values on a calibration curve.
Independent linearity with forced zero:	The linearity defined by referring to the best straight line with forced zero.
Best straight line with forced zero:	That best straight line which goes through the zero point (zero measurand, zero output).
End point linearity:	Linearity referred to the end points.

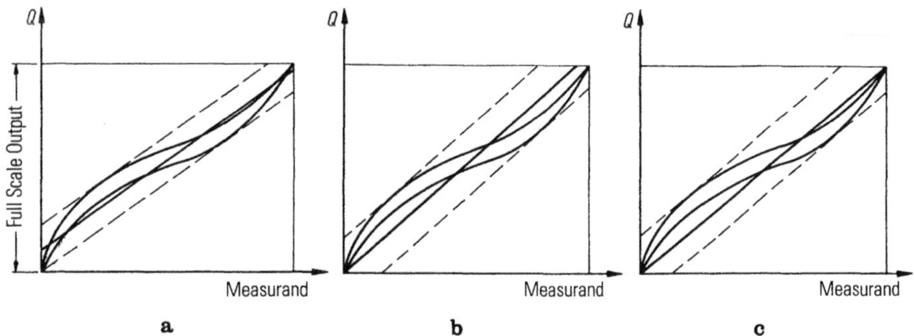

Figure 4.3 Definitions of linearity **a** Independent linearity (the reference straight line is the *best straight line*). **b** Independent .linearity with forced zero (the reference straight line is the *best straight line* that goes through zero). **c** End point linearity (the reference straight line is the *end point line*). Note that here the maximum positive and negative deviations of the calibration curve from the reference straight line are not equal

End points: The outputs at the specified upper and lower limits of the range.
Note: Unless otherwise specified, end points are averaged during anyone calibration.

End-point line: The straight line between the end points.

Full scale output: The algebraic difference between the end points.

When defining the sensitivity, it was already mentioned that it is taken as the slope of that straight line which serves as reference for defining linearity. The best straight line is the best possible approximation of a calibration curve by a straight line. In general this line does not go through zero (zero measurand, zero output) and therefore, it can only be used when that offset can be corrected for in the associated amplifier. In practice a system that gives zero output for zero measurand is preferred, especially in piezoelectric systems where zero is established before each measurement by resetting the charge amplifier (chapter 11). Therefore the independent linearity with forced zero is the most widely used one in practice.

Linearity of a sensor is expressed in percent of full-scale output (%FSO). For the independent linearity with forced zero, the maximum positive and negative deviations of the calibration curve from the best straight line are equal by definition. This linearity is specified as "within ±...%FSO" or also as "≤±...%FSO". Although the latter way of writing "≤±" is not quite correct in the strict mathematical meaning of the sign "≤" it always has been correctly understood in practice to mean "within", because there is no other symbol available – and widely known – which means "within".

Another way of defining linearity is the end point linearity which is useful when the sensor is mainly used near the upper end of the range. However the maximum positive and negative deviations from the reference straight line are no longer equal.

When comparing and judging different sensors regarding their linearity it is very important to know, which definition was used in determining it. The specification "linearity is ±0,8%" alone is meaningless. What is the value the percentage refers to? If it is meant to be the independent linearity, then a sensor with +1,2/0%FSO end point linearity is a better sensor although the numerical value is higher! Therefore, it is essential to know which definition a manufacturer is using in determining the linearity specification of his sensors.

The method of determining the reference straight line for defining linearity by the method of the least sum of the squares of the deviations of the calibration points from the straight line is sometimes proposed. However in calibration the random errors are usually much smaller than the systematic ones because most types of sensors have an inherently characteristic calibration curve which is highly reproducible. Therefore, the least-squares method for defining linearity should be used only when random errors are dominant, justifying a statistical averaging. Besides, independent linearity (with or without forced zero) can easily be determined analytically, too.

There are a number of other methods for defining linearity which are not relevant for piezoelectric sensors [Norton 1969].

4.2.1.3
Influence of Temperature on the Relationship Between Measurand and Output

Operating temperature range: The range of ambient temperatures, given by their extremes, within which the sensor is intended to operate. Within this range of ambient temperature all tolerances specified for temperature error, temperature error band, temperature gradient error, thermal zero shift and thermal sensitivity shift are applicable.

Maximum (minimum) ambient temperature: The value of the highest (lowest) ambient temperature that a sensor can be exposed to, with or without excitation applied, without being damaged or subsequently showing a performance degradation beyond specified tolerances.

Fluid temperature range: The range of temperatures of the measured fluid, when it is not the ambient fluid, within which operation of the sensor is intended.
Note 1: Within this range of fluid temperature all tolerances specified for temperature error, temperature error band, temperature gradient error, thermal zero shift and thermal sensitivity shift are applicable.
Note 2: When a fluid temperature range is not separately specified, it is intended to be the same as the operating temperature range.

Maximum (minimum) fluid temperature:	The value of the highest (lowest) measured-fluid temperature that a sensor can be exposed to, with or without excitation applied, without being damaged or subsequently showing a performance degradation beyond specified tolerances. Note: When a maximum or minimum fluid temperature is not separately specified, it is intended to be the same as any specified maximum or minimum ambient temperature.
Zero-measurand output:	The output of a sensor, under room conditions unless otherwise specified, with nominal excitation and zero measurand applied.
Room conditions:	Ambient environmental conditions, under which sensors must commonly operate, which have been established as follows: a) Temperature: 25 ± 10 °C b) Relative humidity: 90 % or less c) Barometric pressure: 880 ... 1084 mbar Note: Tolerances closer than shown above are frequently specified for sensor calibration and test environments.
Environmental conditions:	Specified external conditions (shock, vibration, temperature, etc.) to which a sensor may be exposed during shipping, storage, handling, and operation.
Operating environmental conditions:	Environmental conditions during exposure to which a sensor must perform in some specified manner.
Zero shift:	A change in zero-measured output over a specified period of time at room conditions.
Thermal zero shift:	The zero shift due to changes of the ambient temperature from room temperature to the specified limits of the operating temperature range.
Sensitivity shift:	A change in the slope of the calibration curve due to a change in sensitivity.
Thermal sensitivity shift:	The sensitivity shift due to changes of the ambient temperature from room temperature to the specified limits of the operating temperature range.
Temperature error:	The maximum change in output, at any measurand value within the specified range, when the sensor temperature is changed from room temperature to the specified limits of the operating temperature range.

Temperature gradient error: The transient deviation in output of a sensor at a given measurand value when the ambient temperature or the measured fluid temperature changes at a specified rate between specified magnitudes.

The operating temperature range is another important property of a sensor, which decides whether a sensor can be used for a given measuring task or not. Within the operating temperature range all specifications are maintained while up to the specified maximum or minimum temperature the specifications must not change beyond specified tolerances. It is of utmost importance that a sensor is never exposed to temperature near or beyond the maximum or minimum temperature specified during operation.

For pressure sensors a fluid temperature may be specified, if it differs from the operating temperature range. The maximum/minimum fluid temperature must be specified, also. During operation all these limits must be respected which can mean that e.g. a cooling adapter has to be used in order to stay within the operating temperature range despite a higher fluid temperature.

The thermal sensitivity shift (Fig. 4.4a) describes the largest sensitivity shift when the sensor is exposed to various temperatures within the operating temperature range. This sensitivity shift is determined after the sensor has reached the new temperature and stabilized and not the transient zero shift occurring during the change in temperature. That shift is covered by the term temperature gradient error.

Piezoelectric sensors can also exhibit a thermal zero shift resulting from the pyroelectric effect inherent in some piezoelectric materials such as tourmaline, piezoelectric ceramics and so on, but not in quartz. Through the pyroelectric (from Greek. pyro = fire) effect the piezoelectric material yields an electric charge in response to a change in temperature only which shifts the zero of the sensor.

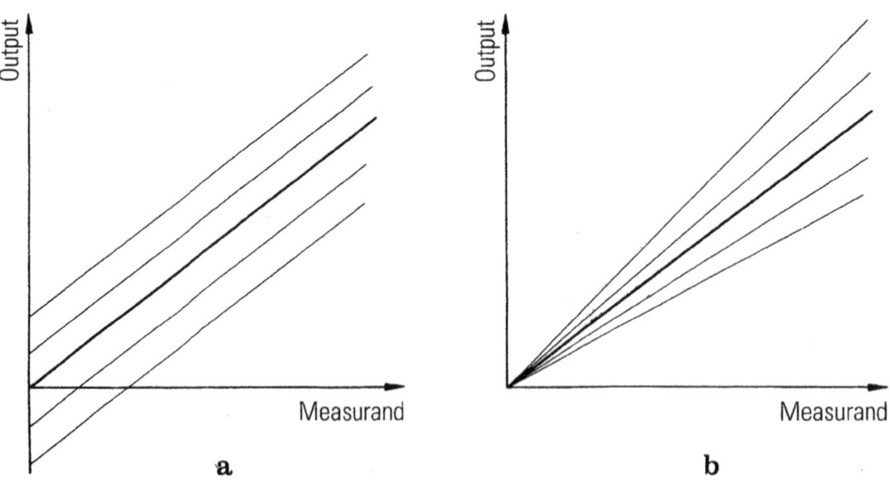

Figure 4.4 Thermal zero shift **a** and thermal sensitivity shift **b**

However many materials (except e.g. tourmaline) that exhibit a pyroelectric effect tend to have a low insulation resistance which results in higher drift of the charge amplifier or results in a short time constant (see chapter 11). This makes it nearly impossible to observe such a thermal zero shift in practice because zero drifts away too fast before the sensor has stabilized at the new temperature. Sensors with quartz element exhibit a pseudo pyroelectric effect which is caused by the different thermal coefficients of expansion of the materials used in the sensor, changing the mechanical preload applied to the quartz element. Yet thermal zero shift is usually of minor importance in piezoelectric sensors because the duration of static measurement is inherently limited and the transient thermal effects are usually much stronger. Thermal zero shift is an important criteria in passive sensors because there it influences directly the measuring error in static measurements over longer time periods and under simultaneous changes in temperature.

However the thermal sensitivity shift (Fig. 4.4b) must be considered in piezoelectric sensors. Thermal sensitivity shift is specified as the biggest change in sensitivity (expressed in percent) when it is determined not at room temperature but at any temperature within the operating temperature range (Fig. 4.5). Because piezoelectric sensors usually have a very wide operating temperature range, rather large values result for thermal sensitivity shift according to this definition. It may be useful to additionally indicate thermal sensitivity shift for a partial operating temperature range mostly used range in practice.

Very often the term "temperature coefficient of sensitivity" – which for good reasons is not contained in the ANSI/ISA-S37.1 standard! – is used in specification sheets. Such a coefficient corresponds to the slope of a tangent on the curve sensitivity versus temperature, therefore such a coefficient is valid only in the vicinity of the tangent point. Specifying a coefficient would make sense only if a linear or nearly linear relationship between sensitivity and temperature existed,

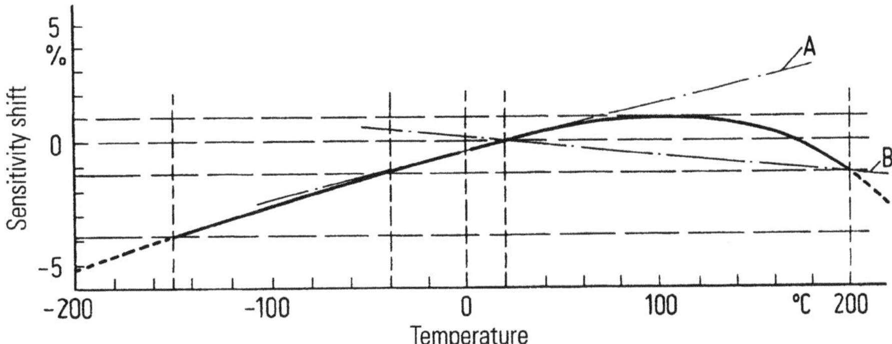

Figure 4.5 Thermal sensitivity shift. The shift is expressed in percent of the sensitivity at 20 °C. Example shown: Operating temperature range: 150...200 °C. Thermal sensitivity shift: 4...1% (in the partial range 40...200 °C: 1,5...1%). **A** Temperature coefficient of the sensitivity at 20 °C: 0,024%/°C. **B** "Average" temperature coefficient of the sensitivity from 20...200 °C: 0,008%/°C

which is not always so with piezoelectric sensors. Completely useless and misleading is the often seen "average temperature coefficient of sensitivity" which corresponds to the slope of a straight line connecting two points on the curve sensitivity versus temperature. Fig. 4.5 clearly illustrates how by using this "method" (and cleverly choosing the two points on the curve...) an average temperature coefficient of nearly zero can be shown in the specifications while in fact it is first positive, then zero, and finally negative!

Thermal sensitivity shift gives only the maximum change over the temperature range looked at. If this is insufficient, then the dependence of the sensitivity on the temperature should be specified in a graph, table or mathematical function of sufficient approximation.

Instead of the various specific errors caused by change in temperature the sensor and other thermal effects, the temperature error (Fig. 4.6) can be specified. The temperature error indicates the maximum change in output at any measurand value within the range when the sensor temperature is varied from room temperature to the limits of the operating temperature range. In other words: the temperature error indicates the deviation of the output from its value determined at room temperature for any temperature and any measurand value within the respective ranges.

Thermal zero shift and sensitivity shift refer to stationary temperature conditions, i.e. they give the differences occurring when the same measurement is made at different temperatures. The temperature gradient error (Fig. 4.6) describe the influence of a rapid change in temperature on the output during a

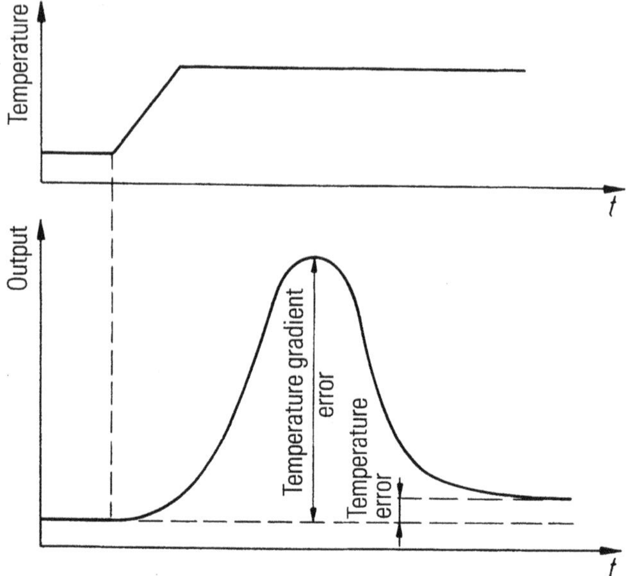

Figure 4.6 Temperature and temperature gradient error

measurement. The values of temperature between which the change takes place as well as the rate of change in temperature must be specified.

4.2.1.4
Influence of Acceleration and Vibration on the Relationship Between Measurand and Output

Acceleration error:	The maximum difference, at any measurand value within the specified range, between output readings taken with and without the application of specified constant acceleration along specified axes. Note: See cross talk (also called transverse sensitivity in the ANSI/ISA-S37.1 standard) when applied to acceleration sensors.
Vibration error:	The maximum change in output, at any measurand value within the specified range, when vibration levels of specified amplitude and range of frequencies are applied to the sensor along specified axes.

Although vibration is by its nature a form of acceleration, a distinction must be made in most cases. A sensor may bear a relatively high constant acceleration while vibration of relatively low amplitude my bring parts of the sensor into resonance at certain frequencies, leading to damage or even destruction of the sensor.

These errors are expressed in %FSO and the test conditions under which they were determined must be specified. They may also be expressed in units of measurand per unit of acceleration , provided that relationship is sufficiently linear over the range looked at. Then the terms acceleration sensitivity or vibration sensitivity can used.

4.2.1.5
Drift, Stability and Repeatability

Drift:	An undesired change in output over a period of time, which change is not a function of the measurand.
Stability:	The ability of a sensor to retain its performance characteristics for a relatively long period of time. Note: Unless otherwise stated, stability is the ability of a sensor to reproduce output readings obtained during its original calibration, at room conditions, for a specified period of time; it is then typically expressed as "within ... percent of full scale output for a period of ... months."
Repeatability:	The ability of a sensor to reproduce output readings when the same measurand value is applied to it

consecutively, under the same conditions, and in the same direction.
Note: Repeatability is expressed as the maximum difference between output readings; it is expressed as "within ... percent of full scale output". Two calibration cycles are used to determine repeatability unless otherwise specified.

When making quasistatic measurements with piezoelectric sensors and using charge amplifiers, drift will always be observed (chapter 11 gives full details). Since drift is not a predictable phenomenon, it is wrong to apply the term drift to the exponential discharge observed in piezoelectric sensors alone or when used with an electrometer amplifier, because that effect is systematic and therefore fully predictable (chapter 11).

The stability of piezoelectric sensors depends mainly on the type of piezoelectric material used as transduction element. Crystals such as quartz, tourmaline and other monocrystals inherently exhibit an outstanding stability while that of piezoelectric ceramics can vary considerably. Depending on their quality, artificial aging and homogeneity they can reach quite good stability or may be at the limit of usefulness. Sensor stability depends also on its design and the materials used in its construction. Especially the diaphragm found in most pressure sensors may be stressed to or even beyond the elastic limit of the material it is made of which can change the stiffness and the active diameter of the diaphragm, resulting in a change in the sensitivity over the time span the sensor is used.

The repeatability of piezoelectric sensors is generally extremely good and is – especially in quartz sensors – within much smaller magnitude than the usual error occurring in calibration.

4.2.2 Dynamic Characteristics

Dynamic characteristics: Those characteristics of a sensor which relate to its response to variations of the measurand with time.

Frequency response: The change with frequency of the output/measurand amplitude ratio (and of the phase difference between output and measurand), for a sinusoidally varying measurand applied to a sensor within a stated range of measurand frequencies.
Note 1: It is usually specified as "within ± ... percent (or ± ... dB) from ... to ... Hz."
Note 2: Frequency response should be referred to a frequency within the specified measurand frequency range and to a specified measurand value.

Natural frequency: The frequency of free (not forced) oscillations of the sensing element of a full assembled sensor.

Note 1: It is also defined as the frequency of an applied sinusoidal measurand at which the sensor output lags the measurand by 90 degrees.
Note 2: Applicable at room temperature unless otherwise specified.

Resonant frequency: The measurand frequency at which the sensor responds with maximum output amplitude.
Note 1: When major amplitude peaks occur at more than one frequency, the lowest of these frequencies is the resonant frequency.
Note 2: A peak is considered major when it has an amplitude at least 1,3 times the amplitude of the frequency to which specified frequency response is referred.

Ringing frequency: The frequency of the oscillatory transient occurring in the sensor output as a result of a step change in measurand.

Ringing period: The period of time during which the amplitude of output oscillations, excited by a step change in measurand, exceed the steady-state output value.
Note: Unless otherwise specified, the ringing period is considered terminated when the output oscillations no longer exceed ten percent of the subsequent steady-state output value.

Overshoot: The amount of output measured beyond the final steady output value, in response to a step change in the measurand.

Rise time: The length of time required for the output of a sensor to rise from a small specified percentage of its final value to a large specified percentage of its final value as a result of a step-change of measurand.
Note: Unless otherwise specified, these percentages are assumed to be 10 and 90 percent of the final value.

The frequency response of a sensor must always be wide enough to cover all frequencies which occur in the measurand to be measured. The lower limit of piezoelectric sensors is always determined by the electronic circuit used to process the sensor output (chapter 11). The upper limit is usually given by the sensor unless an electronic circuit with a lower upper frequency limit is used. In practice the usable upper frequency limit of a piezoelectric sensor is taken as 1/3 of the natural or resonant frequency, whichever is the lower.

The terms natural frequency and resonant frequency are still very often mixed up or even wrongly used. When comparing and selecting sensors, it is very

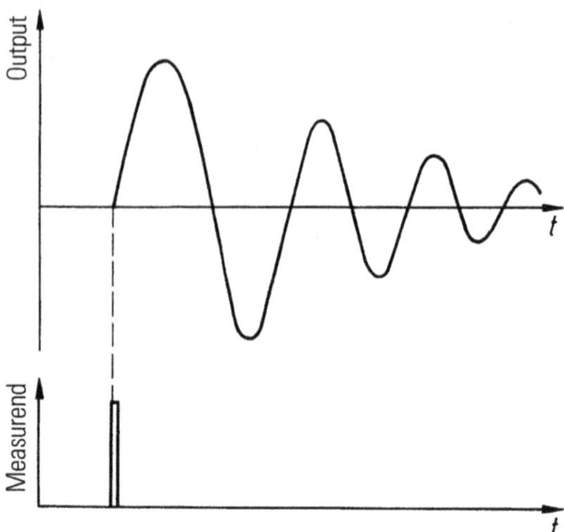

Figure 4.7 Natural frequency. The measurand pulse used to excite the natural frequency should ideally be shorter than a tenth of the period of the natural frequency

important to know exactly, which frequency is specified, what it means and how it was determined.

The natural frequency (Fig. 4.7) is excited by a very short pulse of measurand, whereby the duration of this pulse must be much shorter that the period of the natural frequency. The classic example for demonstrating what natural frequency means is the tuning fork: when struck, it will oscillate at its natural frequency. The amplitude of these oscillations decreases exponentially and therefore, from a correct measurement of natural frequency, the damping can be determined also. Piezoelectric sensors inherently have very high natural frequencies – probably the highest of all commonly used sensor types – and therefore, it is very difficult, sometimes even impossible, to measure correctly their natural frequencies.

For determining the resonant frequency (Fig. 4.8), a sinusoidally varying measurand of constant amplitude and varying frequency is applied to the sensor and the output amplitude recorded against the frequency. A complete measurement of resonant frequency should also include the recording of the phase shift between output and measurand. The ratio of the maximum output amplitude at the resonant frequency divided by the more or less constant amplitude at lower frequencies is called the Q-factor, inherently very high in piezoelectric sensors (usually over 50). Generating a sinusoidally varying measurand of constant amplitude over a frequency range wide enough to exceed the resonant frequencies of piezoelectric sensors is very difficult or – if the amplitude of the measurand has to be of substantial magnitude – just impossible.

The same applies to the ringing frequency (Fig. 4.8) which is often close to the natural frequency (not the resonant frequency). Here too, it is most difficult or even impossible to generate a step change in measurand which is short enough and of sufficient amplitude to properly determine the ringing frequency. Therefore,

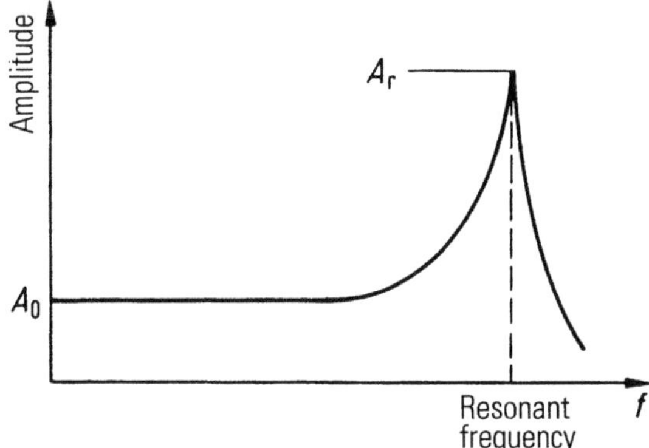

Figure 4.8 Resonant frequency. The sensor is excited by a sinusoidally varying measurand of constant amplitude but varying frequency. From the ratio A/A_0 the rise at resonance and therefore the Q-factor can be determined

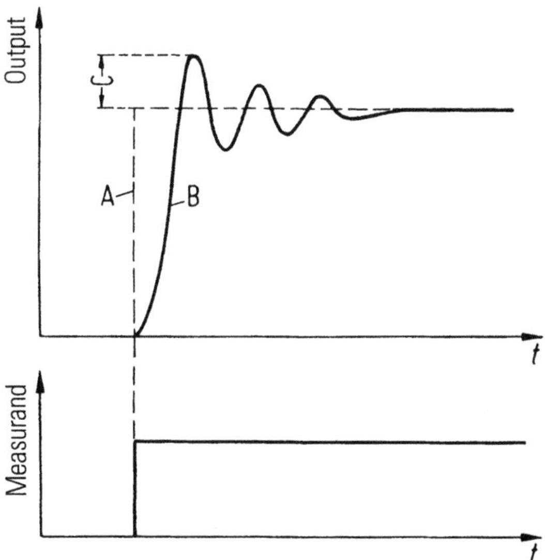

Figure 4.9 Overshoot because of a step-like change in measurand. **A** Output from an ideal sensor. **B** Output from a real sensor. **C** Overshoot, followed by ringing (see ringing frequency)

manufacturers and users of piezoelectric sensors are often resorting to improvised methods which, combined with comparisons and values obtained in practice, still yield usable results. It is recommended that the user inquires in detail on the method(s) used for a specific type of sensor, so realistic comparisons between different makes and subsequent correct interpretation of the measuring results can be made.

When a step change in the measurand occurs, the output of a sensor tends to overshoot (Fig. 4.9). However, step changes in the measurand are rarely rapid enough to provoke overshoot and therefore, overshoot is usually a minor problem in piezoelectric sensors.

For the same reasons, the rise time is also very difficult to measure in piezoelectric sensors. A good rule of thumb, confirmed by long experience and many comparative measurements, is to assume the rise time to be about 25% of the length of the period of the natural frequency of a sensor.

Measuring the dynamic properties of piezoelectric sensors is very difficult, because their advantage of having the widest frequency range and the highest natural frequencies calls for reference sensors and systems which simply do not exist. While for general applications, piezoelectric sensors have a more than adequate frequency response, in research and experimental work at high measurand frequencies, their dynamic characteristics must be carefully analyzed. Also, a close contact with the manufacturer is necessary in order to know which methods were used in determining the published specifications.

4.2.3 Electrical Characteristics

Insulation resistance: The resistance measured between specified insulated portions of a sensor when a specified DC voltage is applied at room conditions unless otherwise stated.

Output impedance: The impedance across the output terminals of a sensor presented by a sensor to the associated external circuitry.

Piezoelectric sensors represent, seen as an electrical device, an active capacitor with a very high insulation resistance (chapter 11). Usually one electrode of this capacitor is connected to the sensor housing. Exceptions are ground-isolated sensors and certain sensors with built-in electronics. In sensors with a transduction element which is fully isolated from the sensor case, the insulation resistance in the strict sense of the definition is understood to be the resistance between one of the electrodes of the transduction element and the sensor case. The output impedance however is measured between the electrodes and has, in piezoelectric sensors, an ohmic resistance in the order of Teraohm (TΩ) and a capacitance in the range of Picofarad (pF). If one electrode is connected to the sensor case – most piezoelectric sensors are designed that way – the insulation resistance becomes identical with the ohmic part of the output impedance. It is general practice to specify the insulation resistance and the capacitance separately for piezoelectric sensors (see chapter 11).

4.2.4 Effects of Sensor Mounting

Mounting error: The error resulting from mechanical deformation of the sensor caused by mounting a sensor and making all measurand and electrical connections.

Strain error: The error resulting from a strain imposed on a surface to which the sensor is mounted.
Note 1: This term is not intended to relate to strain sensors (strain gages).
Note 2: Also see mounting error.

Mounting a sensor on the measured object always entails some mechanical stress and deformation on its case and internal structure containing the transduction element. This can e.g. provoke a change in the sensitivity of the sensor. For sensors which are screwed onto or into the measured object (e.g. pressure and acceleration sensors) often a mounting torque is specified which is equal to the one applied in the calibration setup. Tolerances for flatness may be specified for the surfaces contacting force, strain and acceleration sensors.

Making measurand and electrical connections and placing cables, if done correctly, should not produce any measurable errors in a properly designed sensor.

The strain error must be carefully considered in acceleration sensors because often the structure onto which the sensors is mounted with its base is undergoing changes in strain during the measurement (e.g. flexing oscillations). In practice the term base strain sensitivity (specified as "within $\pm \ldots$ $g/\mu\varepsilon$" is often used for acceleration sensors.

4.2.5 Lifetime of a Sensor

Operating life: The specified minimum length of time over which the specified continuous and intermittent rating of a sensor applies without change in sensor performance beyond specified tolerances.

Cycling life: The specified minimum number of full range excursions or specified partial range excursions over which a sensor will operate as specified without changing its performance beyond specified tolerances.

Continuous rating: The rating applicable to specified operation for a specified uninterrupted length of time.

Intermittent rating: The rating applicable to specified operation for a specified number of time intervals of specified duration; the length of time between these intervals must also be specified.

Storage life: The specified minimum length of time over which a sensor can be exposed to specified storage conditions without changing its performance beyond specified tolerances.

Clear information on the life time characteristics are especially important for sensors used in monitoring and controlling industrial processes. Cycling life is of significance where a high number of cycles of the measurand occur, such as in monitoring vibration levels, cylinder pressure in diesel engines, cavity pressure in injection molding, forces in tabletting presses and so on. A useful parameter is the MTBF (mean time between failures) which should be determined and specified especially for sensors to be used in industrial controlling and monitoring, particularly for sensors working at elevated temperature.

4.2.6
Cross Talk

This term which does not figure in the ANSI/ISA-S37.1 standard [ISA 1982] was introduced by the author in the early sixties in connection with multicomponent force sensors. Such sensors – also called multiaxial force sensors – should ideally give an output signal only at the output which corresponds to the axis along which a force is applied. In practice, small signals also appear at the outputs for the axes normal to the loaded one. The term transverse sensitivity mentioned in the standard – and still used sometimes with acceleration sensors – is not adequate because it implies a ratio of output to some measurand. Cross talk, a term borrowed from the fields of electroacoustics and telecommunications, is defined as follows:

Cross talk: Signal at the output of a sensor which is produced by a measurand acting on the sensor and which is not assigned to that output.

Note: Cross talk is specified as ratio of value of measurand appearing at an output to the value of measurand acting on the sensor in a direction not assigned to that output. If both measurands have the same units, cross talk can be expressed in percent.

Example 1: "Cross talk of F_x to F_y=0,8%" means that if e.g. a force F_x (i.e. a force along the x-axis of the sensor) of say 100 N acts on the sensor, at the output for F_y an output signal is appearing as if a real force F_y=0,8 N had acted on the sensor at the same time.

Example 2: "Cross talk of F_z to T_z=3 Nm/kN" means that e.g. under a force F_z=1 kN the T_z output of the sensors yields an output signal as if a real torque of 3 Nm had acted.

The term cross talk is more universal and can fully replace the term transverse sensitivity for all types of sensors. The practical implications of cross talk are described in detail in chapter 9.

5 Piezoelectric Sensors

5.1 Introduction

Piezoelectric sensors are characterized by having a transduction element made of a piezoelectric material. They are called active sensors because, in principle, no external energy is needed to obtain an output signal.

In most passive sensors, a clear distinction between the sensing element and the transduction element can be made (e.g. in a strain gage force sensor: flexing beam as sensing element, strain gage as transduction element). In piezoelectric sensors, the sensing element and the transduction element are usually one and the same.

In a piezoelectric force sensor the measurand is transmitted directly through solid, rigid metal parts on to the transduction element. The mechanical stress so induced polarizes through the piezoelectric effect the element which yields a proportional electric charge as output.

Pressure sensors usually have a diaphragm (that can be considered as sensing element) which transmits the fluid pressure to the transduction element. In a well designed sensor the effective area of the diaphragm is constant and therefore, the force transmitted to the transduction element is directly proportional to the acting pressure. This force is again converted into a proportional electric charge.

Acceleration sensors are basically force sensors on which a mass – the so-called seismic mass – is attached. When accelerated, this mass (which is the sensing element) , owing to its inertia, exerts a force on the sensor. The seismic mass is constant and the force as well as the corresponding output in the form of electric charge are proportional to the acting acceleration, according Newton's second law.

The measurands pressure and acceleration are therefore measured in the form of force, i.e. the force sensor is the basic type of sensor. Pressure, acceleration and even strain sensors are only particular designs of force sensors.

5.2 Fundamental Observations on Measuring Force

Practically all force sensors have an elastic sensing element whose deformation is dependent on the acting force. In passive types of sensors such as strain gage type, inductive, reluctive, capacitive and so on, this deformation itself must be measured. In order to obtain an adequate sensitivity this sensing element must have a sufficiently high compliance in order to produce a suitable deformation.

Since in piezoelectric force sensors the sensing element is identical with the transduction element that yields the output directly, it is not necessary to measure its deformation as such. This is the key reason why it can be kept smaller by several orders of magnitude than that in passive sensors.

Force sensors should have a rigidity which is as high as possible in order to measure with minimum error as is illustrated in the following example.

Consider a force sensor in the form of a simple spring balance of length x. When loaded, the spring expands whereby the elongation Δx is a measure for the force acting. The relationship between force and elongation is described by the spring constant k.

In the first case (Fig. 5.1), a mass m is suspended on a wire. The tensile force F_s in the wire is $F_s = mg$, with g being the local free fall acceleration. For measuring this force, the spring balance can be used as force sensor. A piece of wire of length x (equal to the length of the unloaded spring balance) is cut from the wire and the spring balance is inserted as the force sensor. The force F_s deforms the spring by Δx which means that the mass m hanging on the wire is now no longer in exactly the same position in space, but lower by Δx. This is the disturbance caused by the inevitable deformation of the force sensor used. However, the measuring error is negligible for all practical purposes (free fall acceleration and barometric pressure, i.e. buoyancy, are only slightly varying with altitude). The relatively large measuring deformation, which is of the order of several cm in a spring balance, has little effect on the result. The force to be measured is $F_s = \Delta x\, k$.

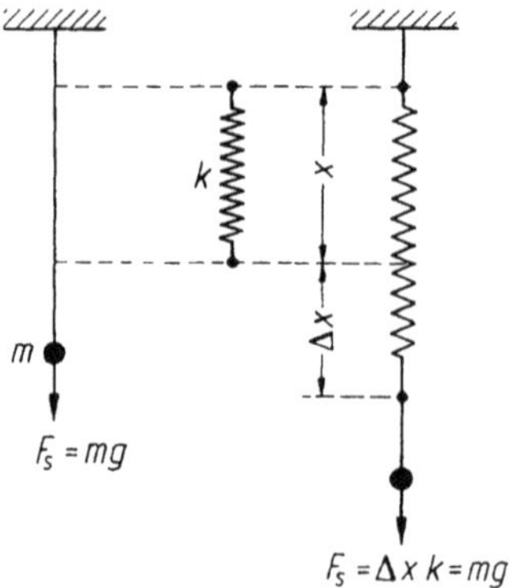

Figure 5.1 Spring scale as simple force sensor

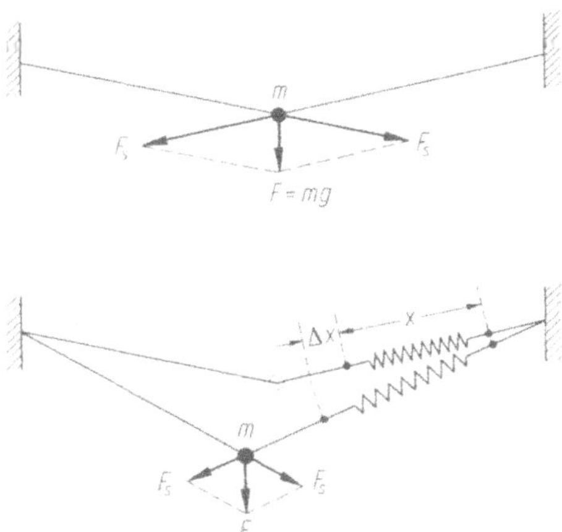

Figure 5.2 Force measurement falsified by an excessive measuring deformation of the sensor

The situation is quite different in the second case (Fig. 5.2). The mass m is now hanging on a wire attached to 2 fixed points. The force $F = mg$ can be resolved into its components F_s which is the tensile force in the wire. Again, the spring balance is used to measure this force by cutting a length from the wire and inserting the spring balance as the force sensor. This time the deformation of the sensor can no longer be ignored. The large measuring deformation of the sensor chosen changes the geometry of the setup in such a way that the actually measured force differs greatly from the force F_s that was intended to be measured originally. The measuring error caused by the measuring deformation of the sensor used renders the measurement completely useless.

This example illustrates dramatically how the rigidity of a force sensor can be of decisive importance in measuring a force, even a static force. In practice the magnitude of the disturbance is rarely visible so clearly, but the same principle applies to all force measurements. As a general rule, force sensors of highest rigidity should be used, unless the influence of the sensor deformation on the measuring error can be analyzed.

The same considerations also apply to measuring pressure, in particular in closed pressure chambers because the rigidity of the transduction element and that of the diaphragm determines the volume change of the sensor which can falsify the result.

The rigidity of the transduction element also greatly influences the natural frequency of a sensor and hence its frequency response. This is of importance in capturing steep pressure rises or shock wave fronts accurately. Sensors with a transduction element of insufficient rigidity will give amplitude errors as well as phase shift errors in response to a step change in the measurand. The resulting low natural frequency will lead to overshoot and ringing.

A general rule is: A higher rigidity of the transduction element leads to a smaller disturbance of the measurement. Piezoelectric sensors have the highest rigidity of practically all generally used sensor types. Of course, the mass between measurand and transduction also influences the natural frequency and must be kept as small as possible.

5.3 Basic Design of a Sensor

All piezoelectric sensors are basically of similar design. The transduction element, which usually acts also as sensing element, consists of one or several piezoelectric elements. The most commonly used piezoelectric materials are single crystals, such as quartz, tourmaline, gallium orthophosphate, crystals from the CGG-group, lithium niobate, lithium tantalate and lithium sulfate, and piezoelectric ceramics (see chapter 3).

There are 4 types of piezoelectric effects: the longitudinal effect, the shear effect, the transverse effect, and the hydrostatic effect. Piezoelectric elements exploiting the longitudinal or the shear effect usually have the form circular, ring-shaped or square plates. In all piezoelectric elements cut for the longitudinal effect and in most elements cut for the shear effect, the faces on which the force is applied and the faces on which the output in form of electric charge appears are the same. In some materials and for special applications, it is possible also to collect the electric charge from faces other than the mechanically loaded ones.

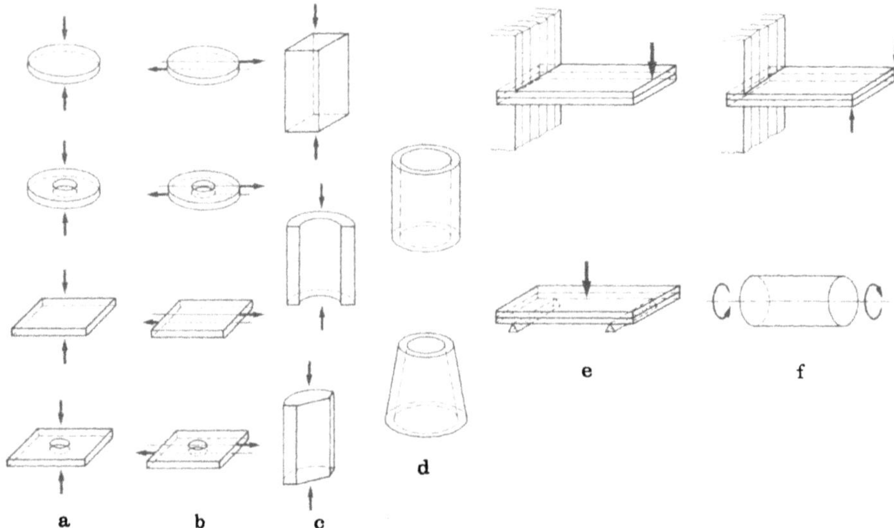

Figure 5.3 Piezoelectric transduction elements **a** plate-shaped elements for the longitudinal effect. **b** plate-shaped elements for the shear effect. **c** rod-shaped elements for the transverse effect. **d** elements in the shape of a hollow cylinder or a truncated cone. Such elements can only be made of piezoelectric ceramics- They can be polarized either radially for the longitudinal effect or in axial direction for the shear effect. **e** bimorph elements as bending beams (exploiting the transverse effect). **f** torsion-sensitive elements (exploiting the shear effect)

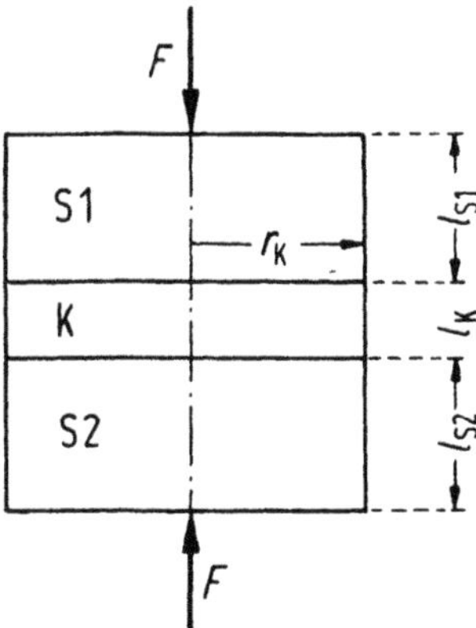

Figure 5.4 Model of a basic force sensor with one piezoelectric element for the longitudinal effect

Elements exploiting the transverse effect are in the shape of rod-shaped in the form of prisms or half-cylinders. In the transverse effect, the force is acting on the end faces and the electric charge always appears on the mechanically unloaded side faces of the element. Therefore these side faces must be fitted with an electrode.

Fig. 5.3 shows examples of practical shapes of piezoelectric elements. A "naked" piezoelectric element can not be used directly for measuring a force. The element must always be fitted between 2 metal blocks which transmit the force to the element and usually serve also as electrodes to collect the electric charge yielded as output. A model of such a piezoelectric force sensor is shown in Fig. 5.4. The piezoelectric element K in the form of a circular plate is located between 2 metal cylinders S1 and S2. The radius r_K of the piezoelectric element is equal to the radius of the 2 cylinders. The thickness of the piezoelectric element is l_K, the height of the cylinders $l_{S1} = l_{S2}$.

Let us assume first that this simple model is loaded by an axial force F, acting normal to the end faces of the cylinders and to the piezoelectric element. We further assume that the force be evenly distributed over the end face of the cylinders. In the piezoelectric element a homogenous state of stress, a homogenous deformation and through the direct piezoelectric effect a homogenous polarization is produced. To simplify we first assume that there is a unixaial stress state in the axial direction, i.e. we neglect the interaction between the cylinders and the piezoelectric elements resulting from the different transverse expansion and the bending deflection of the cylinders. Also we assume that only the compressive force (or later the shear force, too) is transmitted by the cylinders to the piezoelectric element. The thickness of the piezoelectric element is in the direction of the x-axis. The only stress component in the piezoelectric element is

$$T_1 = \frac{F}{\pi r_K^2}. \tag{5.1}$$

Provided the electrodes are short-circuited, the electric field strength in the piezoelectric element is $\boldsymbol{E}=0$ and from the linear piezoelectric equation of state (see [Erhart et al 2002]) we can calculate the electric flux density in the piezoelectric element as

$$D_1 = d_{11}T_1 = \frac{d_{11}F}{\pi r_K^2}. \tag{5.2}$$

The polarization charge Q on the electrodes is obtained by multiplying (5.2) with the area of the electrode. It is πr_K^2 also in our model and therefore

$$Q = d_{11}F. \tag{5.3}$$

The polarization charge which we can e.g. measure with a charge amplifier, is proportional to the acting force. The factor of proportionality is the piezoelectric coefficient d_{11} of the material that the piezoelectric element is made of. It determines the piezoelectric sensitivity of the measuring setup.

If the force distribution across the piezoelectric element is not equal (uneven stress distribution), we obtain the total polarization charge by integrating (5.2) of over the total area of the electrode. Here again, we obtain (5.3).

When the measured force is not normal to the face of the cylinder, it can be resolved into a normal component F_n and a tangent component F_t. The tangent component produces a shear stress (on the lower face of the measuring setup, an equal but opposite reaction force must act in such a way as to prevent the measuring setup from moving). If such a shear stress should not influence the measuring result, it is necessary that

$$d_{15} = d_{16} = 0. \tag{5.4}$$

This condition is e.g. fulfilled by an X-quartz plate. With our measuring setup we then measure only the normal force component F_n.

The condition (5.4) requires that the piezoelectric element is precisely oriented which means that its faces must be exactly normal to the crystallographic x-axis. If this is not so, there will be a measuring error, called cross talk (see 4.2.6). Smallest possible cross talk is of prime importance in multicomponent force sensors (see chapter 6) and the piezoelectric elements must oriented and worked to within a few minutes of angle only.

The fundamental assumption in our highly idealized model is that only a uniaxial stress state is induced in the piezoelectric element by the acting normal force.

In our model this means that the piezoelectric element is completely unconstrained in terms of lateral expansion. In reality the interaction of the surface of the piezoelectric element with the surface of the cylinder results in certain constraints. The problem of compressing an anisotropic piezoelectric plate between isotropic cylinders (usually steel or other metal cylinders) is a general problem of three-dimensional anisotropic elasticity. It appears that up to now an exact general solution has not been found yet. Therefore we can only indicate an approximate solution. We make the following assumptions: a) The contact surfaces remain flat. b) The ratio of the thickness of the piezoelectric plate to its radius is chosen in such a way that the edge effects can be neglected. c) Friction in the contact surfaces is infinitely high, so the deformations of the piezoelectric plate and the cylinders in the contact plane can be considered to be equal. The consequence of these assumptions is that the piezoelectric plate is no longer in a uniaxial stress state.

In addition to the stress T_1 in the piezoelectric plate there are non-homogenous stresses T_2, T_3 and T_4 which influence the polarization charge resulting from the purely compressive load. Their effect changes the effective piezoelectric sensitivity of the piezoelectric plate. This change depends on the elastic properties of the piezoelectric element and the cylinders. For a very thin piezoelectric plate one can safely assume that its lateral strain is entirely imposed by the lateral distortion of the cylinders.

Numeric calculation of the effective piezoelectric sensitivity of a piezoelectric plate that takes into account the interaction with the load-introducing cylinders is quite tedious and is best done by the finite element method [Bertagnolli 1979].

Very simple is only the extremely idealized case where additional stresses are present in the piezoelectric element which suppress all deformations except S_1. Again under the condition that the electrodes are short-circuited, it follows from the piezoelectric state equations that

$$T_1 = c_{11} S_1 \tag{5.5}$$

and

$$D_1 = e_{11} S_1 , \tag{5.6}$$

which means that

$$D_1 = \frac{e_{11}}{c_{11}} T_1 \tag{5.7}$$

and the effective piezoelectric sensitivity is determined by the coefficient e_{11}/c_{11}.

The interaction between the cylinders and the piezoelectric plate has still another significant consequence. Any change in temperature leads to a thermal expansion of the cylinders and the piezoelectric element. Their thermal expansions are in general not the same and that of the piezoelectric plate often even anisotropic. Because of the interaction in the contacting surfaces, stresses are induced that are proportional to the temperature change and which lead to additional piezoelectric polarization charge. The electric output which results in a sensor from a temperature change is sometimes called a false pyroelectric effect (pseudo-pyroelectric effect).

Furthermore, the materials of which the electrodes and the insulation plates sandwiched between cylinders and piezoelectric elements are made of in piezoelectric sensors may contribute to a pseudo-pyroelectric effect (e.g. galvanic effects). In addition the signal resulting from changes in the preload due to temperature change will be superposed on the pseudo-pyroelectric effect. This will be discussed in more details in the context of sensor design.

The pseudo-pyroelectric effect can be suppressed to a large extent by choosing materials of suitable thermal expansion coefficients. Also the thickness of the electrodes and the insulation or compensation plates plays an important role. At the same time, an optimal solution requires a large experimental experience, complemented by simulation techniques. Globally this effect in a sensor is described by the temperature and the temperature gradient error.

For an analogous model of a sensor with a rod-shaped piezoelectric element exploiting the transverse effect, basically similar considerations are valid. Therefore we limit ourselves to calculating the relation between polarization charge and the acting normal force.

We assume that the piezoelectric rod has the shape of a rectangular solid (Fig. 5.5). The length has the direction of the crystallographic y-axis and the width b the direction of the z-axis (such elements are mostly made of quartz). The electrodes are normal to the z-axis and their area is gl. The mechanically loaded areas are ab. The normal force F acting on them induces in the long direction of length the mechanical stress

$$T_2 = \frac{F}{ab}. \tag{5.8}$$

Again assuming only an uniaxial stress state we obtain from the piezoelectric state equations

$$D_1 = d_{12} T_2 \tag{5.9}$$

and the polarization charge on the electrodes is

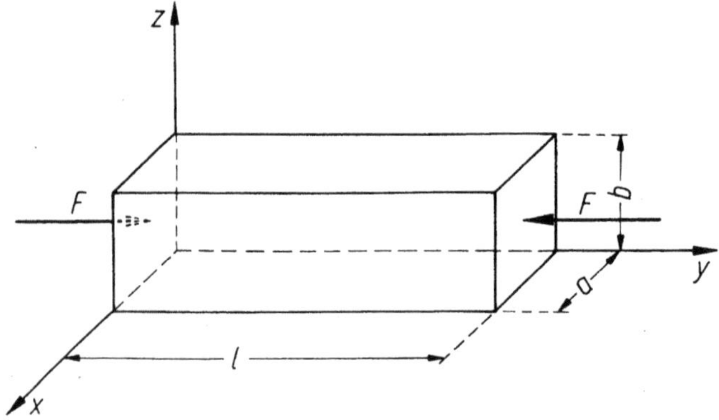

Figure 5.5 Piezoelectric element for the transverse effect

$$Q = D_1 bl = d_{12}\frac{bl}{ab}F = d_{12}\frac{l}{a}F. \tag{5.10}$$

For the transverse effect it is possible to increase the piezoelectric sensitivity by choosing an appropriate ratio l/a ($l>a$) of length to the thickness in x-direction, a possibility which is exploited in high-sensitivity sensors. Mechanical conditions such as the critical buckling length in rod-shaped piezoelectric elements set practical limits for the maximum sensitivity that can be obtained.

The setup in Fig. 5.4 can also serve as model of a sensor for measuring shear forces if a solid connection between the cylinders and the piezoelectric element is assumed. The piezoelectric element must be shear sensitive. With a quartz element this is e.g. fulfilled when the mechanically loaded faces are normal to the y-axis and the force to be measured acts in the x-direction of the piezoelectric element on the cylinder.

The measured force generates – together with the reaction forces that prevent the model from moving or turning – a homogenous stress state

$$T_6 = \frac{F}{\pi r_{\mathrm{K}}^2} \tag{5.11}$$

and therefore an electrostatic flux density

$$D_2 = d_{26}T_6. \tag{5.12}$$

Therefore we obtain for the polarization charge on the mechanically loaded faces, which are normal to the y-axis

$$Q = d_{26}F. \tag{5.13}$$

If a force of arbitrary direction acts on the cylinders S1, our model allows to measure correctly the shear component in the direction of the x-axis of the piezoelectric element only provided that $d_{21}=d_{22}=d_{23}=d_{24}=0$. This condition is fulfilled in a exactly oriented Y-quartz plate. An orientation which is not exact will lead to errors that are described in the chapter on multicomponent force sensors (6.5).

Also in shear-sensitive sensors the significance of the interaction of the piezoelectric element with the shear-transmitting sensor parts must not be overlooked. However, essentially the same considerations as were made for the longitudinal effect apply here, too.

5.4 General Review of Practical Sensor Designs

The simplified and idealized models of sensors just described can not be used for practical application, because it is obvious that e.g. the sensor shown in Fig. 5.4 would simply fall apart. Fixing these parts together by means of an adhesive or by soldering (after coating with metal the piezoelectric element under vacuum) is rarely feasible except in special cases. Therefore the general practice is to mount the piezoelectric element under mechanical preload between the metal parts. The mechanical preload not only assures that the parts are properly fixed together but also has the important purpose to eliminate as much as possible any residual micro-gaps between the contacting faces, a sine qua non for obtaining a good linearity and the very high rigidity necessary for a high natural frequency.

The electric charge yielded by the piezoelectric transduction element is collected by electrodes. Because a piezoelectric sensor is – seen as an electronic component – simply an active capacitor, it is not an electrical contact that is needed between the piezoelectric element (having the function of the dielectric, hence an insulator!) and the electrodes. Usually, one electrode is connected to the sensor housing or formed by one of the force transmitting parts of the housing. while the other one is connected to the center pin of the sensor's coaxial connector. The housing or case of the sensor has 2 functions: it mechanically holds together the inner parts of the sensor and it protects the sensor parts against environmental influences.

Thus, the essential parts of a sensor are: piezoelectric sensing and transduction element, electrodes, preloading elements, housing and connector. Pressure sensors have in addition a diaphragm, which actually is part of the housing and often is also used as preloading element. Acceleration sensors have of course the essential seismic mass. The basic designs of force, pressure and acceleration sensors are illustrated by simplified cross sectional drawings.

Fig. 5.6 shows the cut through a simple force sensor. The transduction element is a piezoelectric plate (*1*), cut for the longitudinal effect. The force F is transmitted by the top plate (2) to the transduction element. The cylindrical wall of the housing (3) which is one part with the base plate (4) is hermetically welded to the top plate and holds the transduction element under mechanical preload. The electrode (5) collects the electric charge yielded by the transduction element and feeds it to the connector (7). An insulating layer (6) is sandwiched between the electrode (5) and the top plate (2) to prevent an electric short-circuit across the transduction element.

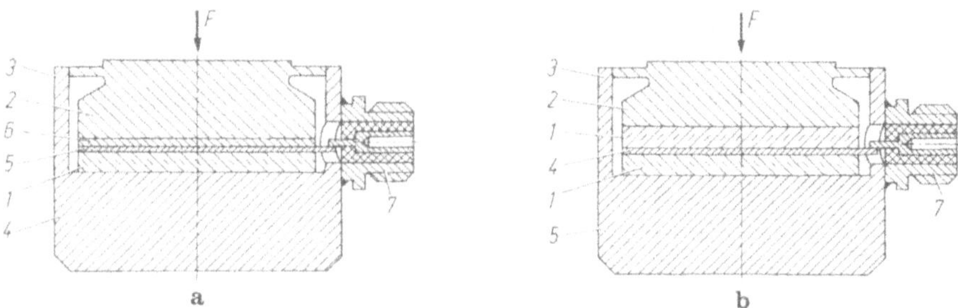

Figure 5.6 Cross-section of a force sensor with one **a** or with 2 **b** piezoelectric plates for the longitudinal effect (after Kistler)

Instead of the insulating layer an identical second piezoelectric plate may be used with its polarity opposite to the other plate. Both plates are then mechanically loaded by the same force and electrically in parallel so that the electric charges yielded by them are summed in the electrode. Such a design has the advantage that no insulating layer is needed which results in a higher sensor rigidity because insulating materials (except e.g. Al_2O_3 and sapphire) usually have a low modulus of elasticity. At the same time, the sensitivity of the sensor is doubled.

The basic design of a pressure sensor with column-type piezoelectric elements exploiting the transverse effect is shown in Fig. 5.7. The fluid pressure p is converted by the diaphragm (*1*) into a proportional force which is transmitted by the front of the preloading sleeve (*2*) and the transfer plate (*3*) onto the piezoelectric elements (*4*), stressing them in their long axis. The electric polarization is in a direction normal to the long axis. The elements (*4*) are mechanically preloaded by the preloading sleeve (*2*). The diaphragm (*1*) is hermetically welded to the sensor housing (*6*). The electric charge yielded through the transverse piezoelectric effect appears on the unloaded side faces of the elements, normal to the axis of polarization. The outer, cylindrical faces of the elements are directly against the inner wall of the preloading sleeve, which acts as the electrode which is connected to the sensor housing. The opposite electrode is formed by the flat face, coated under vacuum with metal, and contacted in several points by the spring-shaped wire electrode (*5*), feeding the electric charge

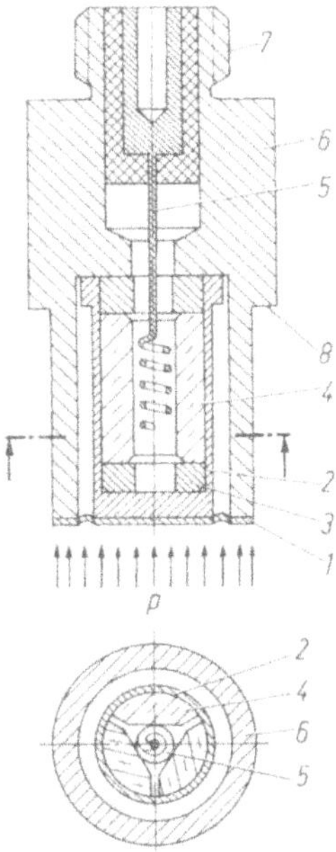

Figure 5.7 Cross-section of a pressure sensor with rod-shaped piezoelectric elements for the transverse effect (after Kistler)

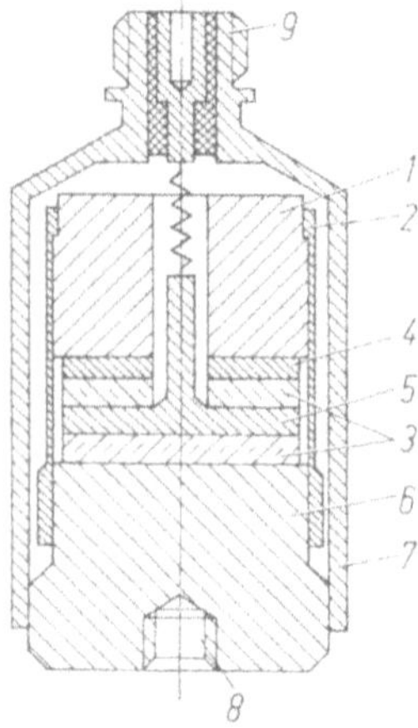

Figure 5.8 Cross-section of an acceleration sensor with piezoelectric plates for the longitudinal effect (after Kistler)

to the center pin of the connector (*7*). The sensor housing (*6*) has a sealing shoulder (*8*).

Pressure sensors are also built with piezoelectric elements cut for the longitudinal effect (Fig. 8.10), especially for achieving high natural frequencies.

Acceleration sensors (Fig. 5.8) always have a mass (*1*) whose inertial force under acceleration is measured by the transduction element (*3*). This mass is often called the seismic mass in analogy with earthquake instrumentation. Here the seismic mass could be considered as the sensing element proper (chapter 9) and therefore be distinguished from the transduction element. The seismic mass (*1*) is usually mechanically preloaded with a preloading sleeve (*2*) on a pair of piezoelectric elements (*3*) cut for the longitudinal effect against the base plate (*6*). The intermediate plate (*4*) made of a material with a suitable thermal expansion serves to compensate the thermal errors because seismic masses are usually made of tungsten carbide (also called heavy metal) whose thermal expansion differs greatly from that of piezoelectric materials. The base plate (*3*) is hermetically welded to the housing (*3*).

These examples give a general overview. More details and other designs are described in chapters 6 through 10.

Charge amplifiers are by far the most commonly used amplifier type with piezoelectric sensors. Almost all commercially available charge amplifiers give a negative output voltage in response to a positive electric charge at their input and vice versa, i.e. they reverse the polarity of the signal. Therefore the piezoelectric elements are mounted in the sensor in such a way that a positive measurand yields a negative electric charge as the sensor output. Processed by a charge amplifier this results in a positive output voltage at its output in response to a positive measurand acting on the connected sensor.

5.5
Components of a Sensor

5.5.1
Transduction Elements

For transduction elements either piezoelectric single crystals or piezoelectric ceramics are used. There is a wide choice of such materials, yet only relatively few have become of importance in practice. Quartz is by far the most used single crystal in sensors. Tourmaline and lithium niobate are found in sensors for measuring at high temperatures, mainly in vibration sensors. More recently single crystals from gallium orthophosphate and the group of CGG are used, too, especially in pressure and force sensors for high temperatures.

In the field of man-made piezoelectric ceramics, numerous materials have been developed by the various manufacturers. They are mainly based on barium titanite, lead circonate, lead niobate and lead metaniobate. The exact composition of these materials is usually not known because it is the know how of the respective manufacturers and therefore kept secret for commercial reasons.

5.5.1.1
Quartz

In the early days, naturally grown quartz was the source for transduction elements made of α-quartz. Natural quartz – a mineral of great beauty – often has small defects such as inclusions, twinned regions, cross-growth, fractures and so on yet its price is high because it is sought after by mineral collectors. Only about 20 % of a natural quartz is suitable for cutting into transduction elements.

Therefore, artificially grown quartz has now replaced the natural one especially since it has become possible to grow ingots to over 60 × 60 mm cross-section, sufficient for even large force sensors. The artificially grown quartz ingot has the shape of a prism (Fig. 3.5). A reference flat is ground on a long side of the ingot and a first orientation done under polarized light. For the precise orientation of the crystallographic axes an x-ray goniometer (Fig. 5.9) is used [Heising 1946; Jost 1975; Raaz 1975].

The quartz ingot is then cut into thin wafers with a special reciprocating gang saw (Fig. 5.10). The saw blades made of hardened steel are toothless and the cutting action is achieved by an oil emulsion of corundum powder. Corundum has the hardness 9 on the Mohs' scale while quartz and steel are of hardness 7. Therefore not only the quartz is cut but the saw blades are worn down too, and can usually be used once only, sometimes twice by flipping them over for the second cut. Using ultrasonic methods circular or ring-shaped transduction elements are then cut from these wafers (Figs. 3.5 , 5.3 and 5.11).

The orientation of the crystallographic axes are verified and, if necessary, corrected to be within the required tolerances. Various steps follow such as finishing the surfaces by lapping, polishing the edges and ultrasonic cleaning for

Figure 5.9 X-ray goniometer for orienting of quartz ingots (Courtesy of Kistler)

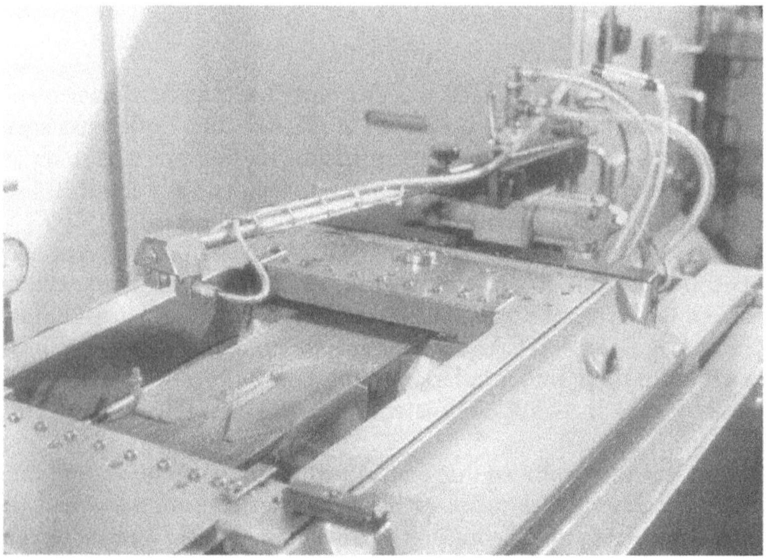

Figure 5.10 Gang saw for cutting single crystals (Courtesy of Kistler)

insuring a high insulation resistance before the element is ready to be mounted into a sensor. In quartz elements cut for the transverse effect 1 or 2 faces on which the electric charge appears are coated under vacuum with metal to form an electrode that can be contacted e.g. with a touching wire electrode.

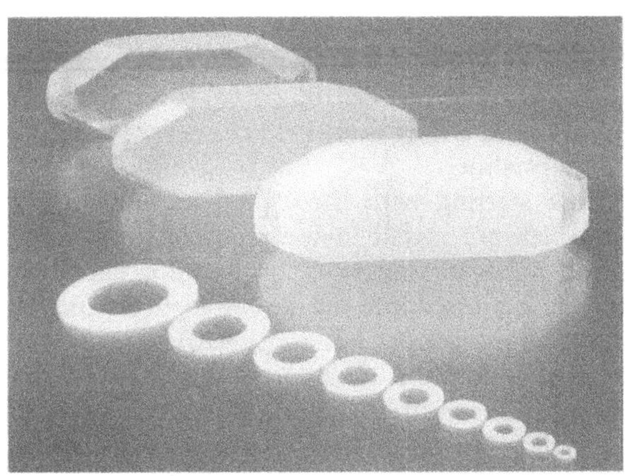

Figure 5.11 From the quartz ingot to the finished tranduction element (Courtesy of Kistler)

5.5.1.2
Tourmaline

Tourmaline is very difficult to grow artificially, especially of sufficiently large size and in commercially viable quantities. Natural tourmaline continues to be used although the great variation found in tourmaline depending on its origin leads to a low yield. The methods for making transduction elements from tourmaline are similar to the ones used for quartz. Since tourmaline is a natural crystal, wafer cut from it must be inspected very thoroughly for inclusions and cracks. Especially when using black (opaque) tourmaline, only very thin wafers can be inspected reliably by just visual methods.

5.5.1.3
Crystals of the CGG-Group and Gallium Orthophosphate

These synthetic crystals are grown by some manufacturers, who have invested into their own crystal growing equipment and subsequently built up their own specific and proprietary know how. Primarily, these crystals are used by the respective manufacturers for use in their own sensor production. Gallium orthophosphate is now available on the market from the AVL company while crystals of the CGG-Group so far are grown and used exclusively by the Kistler company only. The methods for making transduction elements from these crystals are similar to the ones used for quartz.

5.5.1.4
Piezoelectric Ceramics

While the properties of most components used in current ceramics are well known the exact composition and the subsequent processing (especially aging and method of polarization) of commercially exploited ceramics is usually kept as a trade secret by the respective manufacturers.

Examples of such materials are the Piezite types P-1, P-4, P-6, P-10 and P14 by Endevco, MT 8, PZ 23, MT 40 and PZ 45 by Brüel & Kjær, as well as the PZT types (chapter 3). It should be noted that MT 100 and PZ 100 (Brüel & Kjær), Piezite P-2 (Endevco) and V 1 (Vibrometer) are quartz; Piezite P-15 (Endevco) is lithium niobate; and V 2 (Vibrometer) is tourmaline.

Piezoelectric ceramics are made starting with the ingredient materials in powder or fine granulate from which are mixed in the required proportions, sintered at high temperatures and pulverized again. Then bonding agents are added, and the granulate so obtained is pressed into the desired form and fired in a oven under closely controlled conditions (see 3.7.1 and Fig. 3.20).

The elements are fine-machined and fitted with electrodes. The ceramic material must be polarized to become piezoelectric. This is achieved by heating the element to a closely controlled high temperature and applying a high voltage across the electrodes for a certain time. Finally, the elements are "aged" by subjecting them to various temperature cycles which leads to a higher stability of the piezoelectric properties of the material.

5.5.2 Electrodes

Piezoelectric sensors are active capacitors with the transduction element acting as "active" dielectric. The electrodes have the function of collecting the electric charge appearing on specific surfaces of the transduction element and feeding it to the connector of the sensor. In "single-ended" designs, one of the electrodes is formed by a part of the sensor housing – this can be the preloading sleeve, the seismic mass or directly the wall of the sensor housing – and therefore connected to the mass of the sensor. A thin metal foil or a vacuum-deposited metal film serves as the second electrode.

The always different thermal expansion characteristics of transduction material and electrode does influence the thermal characteristics of the sensor and often it is necessary to use electrodes of greater thickness and of special alloys to minimize thermal influence.

The faces of the transduction element that yield the electric charge must be covered entirely by the electrode to collect all elecgric charge and to prevent that high static voltage may result on an uncovered surface, which might result in a spark over.

In all sensors which have piezoelectric elements on which the electric charge yielded appears on the mechanically loaded surfaces (longitudinal and shear effects) the electrodes are pressed against the piezoelectric elements by the always present preload. A good mechanical contact is necessary to minimize or eliminate any micro-gaps between the parts, a sine qua non for a high rigidity of the sensor. Also it assures that the capacitance of the sensor is clearly defined.

Transduction elements on which the electric charge appears on the mechanically not loaded surfaces (transverse effect) must be fitted with electrodes in the form of vacuum-deposited metal films. This insures that all electric charge is collected which then can be fed with a wire – either soldered to the film or just mechanically touching it – to the connector of the sensor.

5.5.3 Insulating Materials

In piezoelectric sensors, insulating materials must meet extreme requirements. The insulation resistance must be over 10 TΩ at thickness of less than 1 mm and surfaces of a few mm^2. This value should drop as little as possible with increasing temperature and/or humidity. The mechanical compressive strength should be at least about 200 N/mm^2.

In practice, materials such as PTFE (e.g. Teflon®) and Kapton® are mostly used for applications at temperatures below 200 °C because their insulation is better than 1 PΩm. Kapton® has a higher mechanical strength which makes it the preferred choice when the insulator is between electrode and preloading element, i.e. in the direct flow of the force through the sensor. PTFE tends to start to creep under increasing mechanical stress and its coefficient of friction drops, which could lead to shifting of the parts against each other. This also restricts the capacity of transmitting shear forces in sensors using the shear effect. However, PTFE is an ideal insulator for connectors and cables

Also there is a number of potting compounds on the market which can be used as insulator in sensors.

In all insulating materials the insulation resistance drops with increasing temperature. Also their mechanical properties are impaired (softening, crumbling). The useful temperature range of PTFE goes up to about 200 ... 250 °C, that of potting compounds to about 150 ... 200 °C while Kapton® can be used up to around 300 °C. For higher temperature up to about 600 °C only ceramic insulators can be used. Still even in ceramic insulators the insulation drops with increasing temperature. In metal jacket cables, aluminum and magnesium oxides as well as mineral fibers and mica have been used successfully, too.

5.5.4 Preloading Elements

An ideal mechanical preload should be as elastic as possible, i.e. much softer than the transduction element. This keeps the force shunt small and also will give a preload which varies only a little over the range of elastic deformation of the transduction element. From the mechanical analysis given in chapter 6.2.3 follows that preloading elements should be long and have a small cross section, i.e. be as elastic as possible. Therefore high strength steels with elastic limits of over 1 kN/mm^2 and also special copper-beryllium alloys are used.

The most common types of preloading elements are the preloading sleeve (e.g. Fig. 5.8) and the preloading bolt (e.g. Fig. 6.23). Sometimes, part of or even the entire sensor housing is used as preloading element (e.g. Fig. 9.9a).

5.5.5 Sensor Housings

The sensor housing has several functions. In the first place it must protect against dirt and humidity in order to maintain the high insulation resistance. The housing also serves as an electric shield to allow measuring uninfluenced by electric fields. Sometimes the housing must resist external pressure such as in acceleration sensors used in pressurized fluids.

These requirements can only be met by a hermetically sealed, water and gas proof housing. Therefore the housing parts are usually welded tight together. The housing material must be sufficiently corrosion resistant. Although the housing itself is easily sealed by welding, the connector presents a problem. Connectors with PTFE insulation are never really hermetically tight (see 5.5.6). In practice this means that tightness is only achieved after connecting the cable. Often, the connection is additionally sealed with a shrink sleeve or a suitable potting compound. Sensors having ceramic-insulated connectors (metal-to-ceramic bond) or with welded-on metal cables are completely tight.

In force sensors parts of the housing are used to transmit the force to be measured onto the transduction element. Also the housing often directly serves as a preloading element and must therefore have suitable elastic properties. Loading faces and mounting threads are part of the housing. Therefore the housing material must of of sufficient mechanical strength. The same applies to the sealing shoulders of pressure sensors.

5.5.6 Connectors

Usually piezoelectric sensors need only a single-pin coaxial connector. In sensors with a transduction element electrically isolated from the housing (so-called ground-isolated sensors) connectors with 2 pins within the shield or coaxial connectors with 2 concentric shields (so-called biaxial connectors) are used.

The most widely used connector is the "Microdot" connector with a 10-32 UNF thread. It was originally introduced by the Microdot company in California which also gave it the name. Numerous manufacturers have developed their own variations of this connector family and usually they are compatible amongst each other. Still it is wise to check because pin length and tolerance of the diameter may give problems or even lead to damage e.g. to ceramic-insulated connectors on sensors when different makes are mixed.

For larger sensors, dynamometers and force plates, BNC and TNC type connectors are preferred because they are much more rugged than the "Microdot" connector. Moreover TNC connectors are water tight.

Multicomponent sensors must have either several single-pin or one multi-pin connector (such as e.g. made by Fischer® or Lemo® in Switzerland). Again these connectors are modified standard versions, with PTFE insulation and special potting compounds, where needed.

The generally used insulation material is PTFE because it has excellent electrical properties and can easily be machined. The major drawback is that such connectors are not tight because PTFE can not be bonded to the metal housing of the connector. PTFE flows under pressure which means that even if the PTFE insulator is pressed into the connector housing, the preloading so obtained will be gone within a short time. Therefore such connectors should always be sealed with the protective caps provided by the manufacturer to prevent the sensor from "inhaling" humidity through its connector due to changes in temperature and ambient pressure when no cable is connected.

For reliable continuous operation PTFE can only be used up to about 200 to 240 °C. For higher temperature connectors with glass or ceramic insulators have to be used. Such connectors are tight but their insulation resistance drops to about 1 to 100 GΩ at higher temperature.

Instead of a connector on the sensor the cable can be attached directly to the sensor (so-called "integrated cable"). This is of advantage especially in small sensors where there is little space for a connector. Integrated cables are also used in sensors designed for high temperature, they are usually welded-on metal cables.

With more and more piezoelectric sensors being used in industrial process control, manufacturers have developed special design connectors and cables to meet the degree of protection IP67, IP69 and so on.

6 Force and Torque Sensors

6.1
Quantity, Units of Measurement, and Coordinate Systems

The SI (Système International) unit of measurement for force F is the „Newton“ (N), which is one of the 7 base units of that system. In practice, large forces are usually expressed in kN (kilonewton) or MN (meganewton) while for small forces, the mN (millinewton) is used.

In physics and stress analysis a force or stress is considered positive when it produces a tension (pulling action), in the field of measurement – especially with piezoelectric sensors – a force which compresses the sensor is usually considered as positive.

Torque and moment is measured in Nm (Newton-meter; not mN, which could be mistaken for millinewton). Also used are kNm (kilonewton-meter) for large, Ncm (Newton-centimeter) for small torques and moments.

The quantities torque and moment are often confounded, yet they are of distinct different nature. A torque T is defined as the moment of a couple, i.e. the twisting or torsional moment caused by 2 parallel forces of equal magnitude but opposite direction and not acting on the same line. Therefore, a torque has no resultant external force because the 2 forces of the couple cancel each other and only the torsional moment results. Also a torque does not have a "point of application". In contrast, a moment M, defined as the moment of force about a point which is equal to the product of the magnitude of the force F multiplied by the distance d of the line of action of the force from the point, clearly has a resultant external force. A moment of force M can always be replaced by a torque T and a force F of equal magnitude and direction going through the point of reference.

In multicomponent sensors and measuring systems, usually a positive (right-handed) Cartesian system of coordinates is used. It is generally oriented in such a way that the positive z axis is in the direction of a compression force on the sensor. This coordinate system must not be confounded with the crystallographic coordinate system of the piezoelectric materials.

6.2 Force Sensors

6.2.1 Sensors for Compression Force

The basic design has already been described in chapter 5.4, Fig. 5.7. Derived from that design a very useful type of force sensor is the so-called load washer (Fig. 6.1). The ring-shaped base plate (1) with thin cylindrical walls is machined from one piece of steel. 2 ring-shaped quartz plates (3) are held under a small preload by the top plate (2) because the top plate is welded on under a closely controlled preload. The output signal is picked up by the electrode (4) placed between the 2 quartz plates and fed to the connector (5).

The quartz plates are cut for the longitudinal effect, i.e. normal to the crystallographic x axis. The nominal force sensitivity is 2,30 pC/N per x-cut quartz plate (see 3.2.2). The cylindrical walls and the diaphragm-like bridge to the load-bearing parts of the sensors form a mechanical shunt to the transduction element, i.e. a small part of the force to be measured goes through them and is not measured. Also the mechanical interaction between the quartz element and the force-transmitting steel plate, each of them having a different modulus of elasticity, influences the sensitivity because lateral expansion is not free. The typical sensitivity of a x-cut quartz plate so mounted is about 2 pC/N. The 2 quartz plates are mounted with their positive crystallographic x-axis in opposite directions which means that they are electrically in parallel, mechanically in series. Therefore such a sensor has a typical sensitivity of –4 pC/N. The reason for the negative sign is given in 11.5.1.

Load washers are available with ranges from a few kN up to over 1 MN (Fig. 6.2). The rigidity (mechanical compliance) of these sensors is in the range of 1 kN/µm (1 µm/kN) for the smaller ones and up to about 100 kN/µm (0,01 µm/kN) for the larger ones. However the threshold of all these sensors is the same, in the order of a few mN, because the sensitivity of all these sensors is the same. Therefore, if the rigidity of the sensor is critical (see chapter 2.2) a sensor with a range larger than really required can be chosen, which offers the needed rigidity. Thanks to the same low threshold, small forces can still be measured with the same accuracy as if a sensor with a smaller range were used. Also the sensor with the larger range will result in a setup with a higher natural frequency. Yet

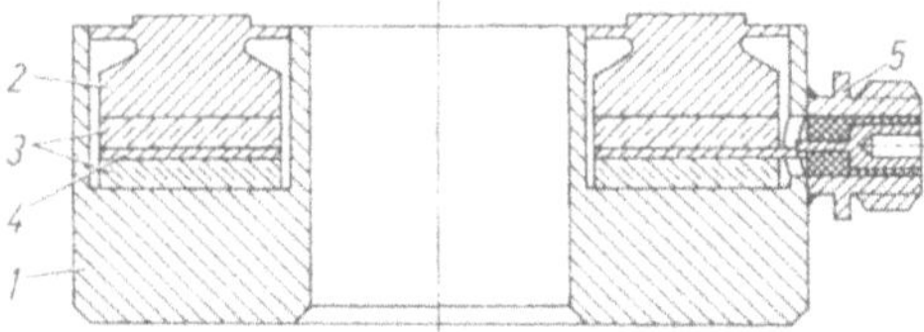

Figure 6.1 Force sensor in the shape of a washer, also called load washer (after Kistler)

Figure 6.2 Load washers with ranges from 0...7,5kN at 10mm external diameter up to 0...1,2MN at 245mm external diameter (Courtesy of Kistler)

another advantage of being able to chose a sensor with too large a range is that when the maximum peak force is not known, the sensor with the larger range provides a higher overload margin without compromising the measurement of a small force.

6.2.2 Linear Compression Force Sensors

A novel type of force sensor specifically developed for WIM (weigh-in-motion) applications illustrates a particular variation in the basic quartz load washer design. This sensor (Fig. 6.3) has a load-bearing area of only 50mm width and e.g. of 750 or 1000mm length. Figure 6.4 shows the cross-section of that sensor which can easily be inserted in a slot previously prepared in the pavement of a road. Quartz elements of rectangular shape, a strip-shaped electrode and a strip of

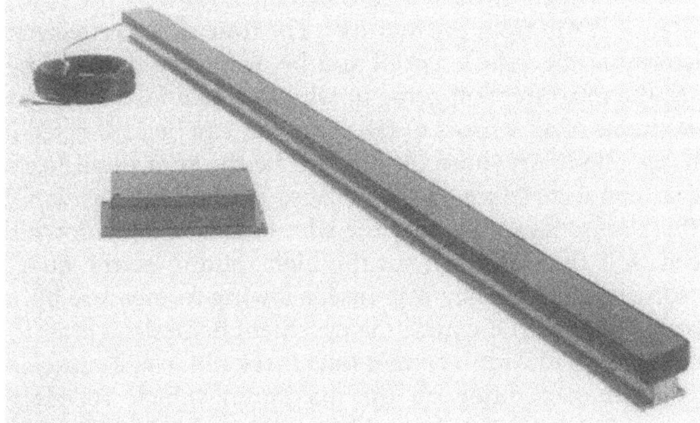

Figure 6.3 Linear quartz force sensor for WIM (weigh-in-motion) applications (Courtesy of Kistler)

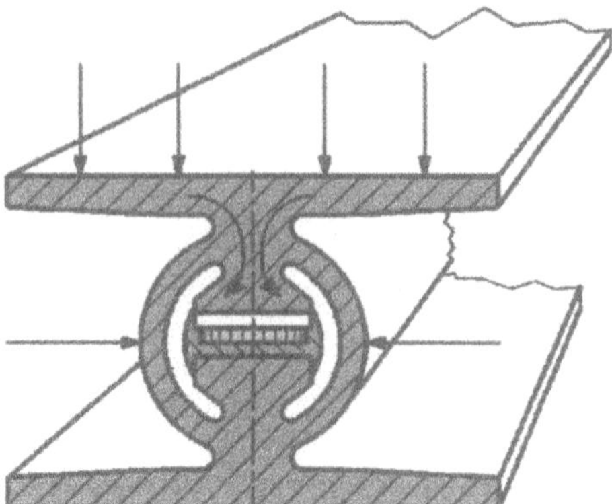

Figure 6.4 Cross section of the linear force sensor shown in Fig. 6.3. The quartz elements are in the center of the profile (Courtesy of Kistler)

Kapton® foil serving as insulating material are sandwiched in the central gap of a drawn light alloy profile. For assembly, the cylindrical part of the profile is laterally squeezed by a controlled force in a transverse direction which will slightly widen the central gap and allow to insert the quartz elements, electrode and insulator. When taking off the squeezing force the profile will hold the inserted parts securely in place under a certain mechanical preload, similar to the principle of a load washer (see 6.2.1). Provided the profile and the parts have dimensions of the required close tolerances, the sensor will not only exhibit an excellent linearity but also a constant force sensitivity along its long axis.

Vehicles traveling across the sensor produce signals from which the wheel or axle load of the vehicle can be determined by integrating the area of the force signal and taking into account the vehicle speed. Usually, a second sensor – or row of several aligned sensors to cover the whole width of the lane or the road – is installed e.g. at a distance of 4 m from the first row. The time shift between the 2 signals allows to determine the vehicle speed and by averaging the readings from both sensor rows, the measurement error in wheel or axle load is further reduced. The quartz elements of all sensors of the same row can be connected in parallel to a single charge amplifier whose output will be the sum of all forces acting on all sensors so connected (see 11.6). Because of the high insulation resistance of quartz, these sensors can measure the wheel loads reliably down to very slow vehicle speed (less than 1 km/h), yet the high rigidity of the quartz elements give the sensor a wide frequency response, allowing to measure up to vehicle speeds far beyond 100 km/h. Of course the pavement before and after the sensors must be flat and smooth to within certain tolerances and over a sufficient distance (some 30 to 100 m, depending on the maximum vehicle speed) to minimize measuring errors due to vertical vehicle bounce [Calderara 1996; COST 1999].

6.2.3
Measuring Tension Force with Preloaded Force Sensors

Piezoelectric transduction elements will give a correct output when loaded by a tension force. However it is usually not practical to use a piezoelectric material under tension because such materials are brittle and it is very difficult to transmit a tension force to them. Therefore the technique of preloading is used, i.e. a piezoelectric force sensor is mechanically preloaded and tension forces can then be measured even so the piezoelectric transduction element always remains under mechanical compression.

An ideal preload would be infinitely elastic and therefore not form any mechanical force shunt, i.e. the entire force to be measured would pass through the sensor. This can be achieved with a so-called preload with dead weight. An anchor point for a cable can be taken as an example (Fig. 6.5). A force sensor such as a load washer is mounted under the anchor point and preloaded by the mass m which serves as dead weight. The preloading force is $F_V = mg$ and – as there is no mechanical force shunt – remains constant as the preloading force results only from gravity. The sensitivity of the sensor for tension force F_s is the same as for compression force as e.g. found by calibrating under a hydraulic press (see 6.8). F_V is the only force acting on the sensor as long as the cable traction F_s is zero.

If a force F_s is applied, the force $\Delta F_A = F_s - F_V$ acts on the sensor (Fig. 6.7a). Since F_V is constant and the corresponding output was eliminated (tared) before the measurement by simply resetting the charge amplifier (see 11.5.3) the load washer now measures as intended only the force F_s because $\Delta F_V = 0$ and therefore $\Delta F_A = F_s$. Naturally the measurement will stop when $F_s > F_V$ because then the anchor point including the mass m would simply start to move away – the preload being infinitely elastic! The essential points in this example are, as long as $F_s < F_V$, that the preloading force F_V is constant, independent of the acting force F_s (property of

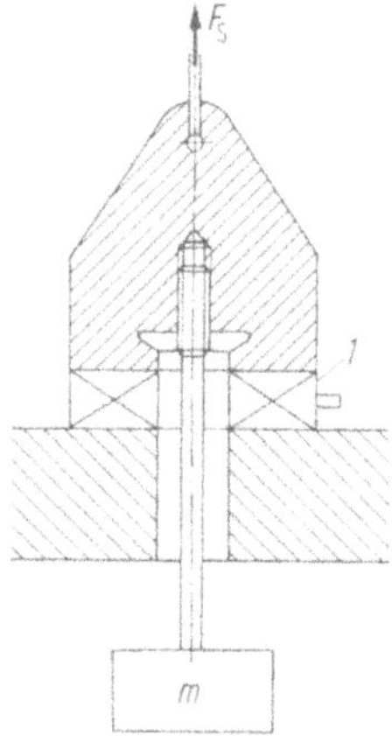

Figure 6.5 Ideal preload with a "dead weight" illustrated with an anchor point of a cable. F_s tension force in the cable, *1* sensor, *m* mass acting as "dead weight"

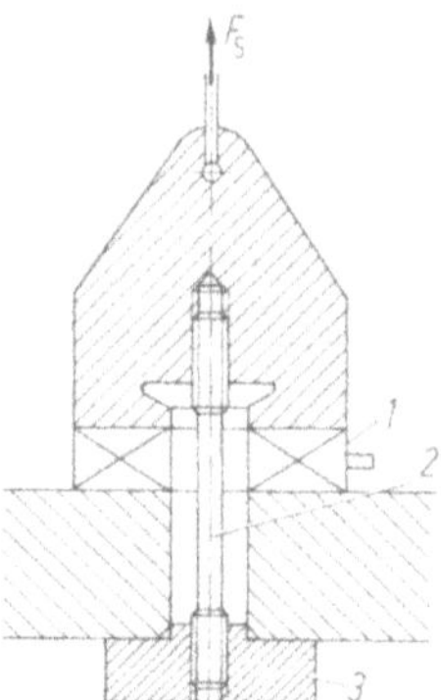

Figure 6.6 Practically feasible preload with a preloading bolt for a cable anchor point. F_s tension force in the cable, *1* sensor, *2* preloading bolt, *3* preloading nut

preload with dead weight) and that the total force F_S is measured by the sensor as there is no mechanical force shunt created by this type of preloading.

This "ideal" preload with a dead weight is however not feasible in most practical applications. Therefore the force sensor is preloaded e.g. with a preloading bolt against the base plate (Fig. 6.6). This kind of preload is not infinitely elastic as is the preload with dead weight. The elasticity of the preloading bolt is now determinant for the mechanical property of the setup. The bolt forms a force shunt, i.e. part of the force F_S to be measured passes – unmeasured – through the bolt. Therefore the sensitivity of the so mounted sensor is now lower and that sensitivity must be determined by calibration after the preload has been applied by tightening the bolt accordingly. For measuring the applied preload during tightening the bolt, the force sensor can be used directly, using its sensitivity obtained during calibration for compression force. In order to have a small force shunt a preloading bolt of small cross section and sufficient length should be used which calls for high strength materials such as special alloy steels (e.g. 17-4 PH, RHF-32, RHF-33, 1.4021 and 1.4045) or special copper-beryllium alloys, having elastic limits higher than 1 kN/mm^2. In a properly designed preloading setup the force shunt can be kept around 5...15%. In all cases, after installing and preloading a force sensor, its sensitivity must be determined again by calibration.

A preloading bolt of high elasticity – hence a small force shunt – is desirable for 2 reasons: The sensitivity of the sensor is reduced only a little through preloading and the preload acting on the sensor will only vary little over the measuring range of the sensor. This latter point is very important in 3-component force sensors because there shear forces are transmitted by friction and a small change in preload over the range of compression force will affect the available range for shear force much less.

The diagrams in Fig. 6.7 clearly illustrate this. Thanks to the extremely high rigidity k_A of piezoelectric sensor the measuring deformation $\delta = F_V/k_A$ is correspondingly small when the preload F_V is applied. In a good preloading setup

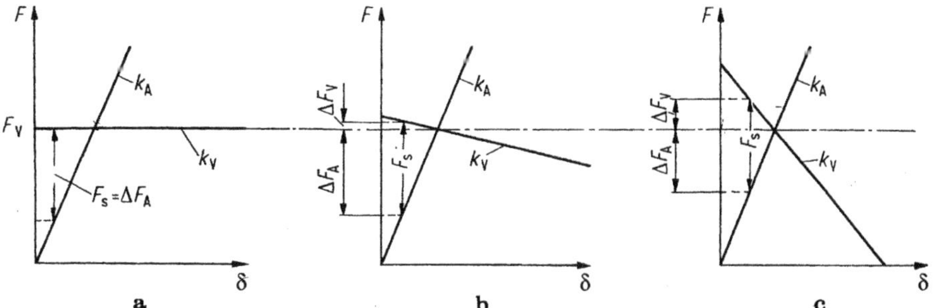

Figure 6.7 Force versus deformation for **a** an ideal preload with "dead weight", **b** a preload with thin, elastic preloading bolt (good solution) and **c** a preload with a thick, inelastic preloading bolt (poor solution). F force (positive for compression) acting on the sensor, δ deformation, F_s tension force in the cable, F_V preloading force, ΔF_V change in the preloading force due to the action of F_s, ΔF_A change in the force acting on the sensor due to the action of F_s, k_A spring constant of the sensor, k_V spring constant of the preload

the elasticity of the bolt is much higher than that of the sensor and the spring constant of the bolt is much lower than that of the sensor ($k_V \ll k_A$). Then the change in the preloading force F_V is very small so it can be assumed to be almost constant over the measuring deformation δ ($\Delta F_V = \delta k_V$) of the sensor. In practice ΔF_V is usually below $0{,}1F_V$. If the force F_s to be measured is acting again it will be measured by the sensor as $\Delta F_A = F_s - \Delta F_V$. Moreover the influence of ΔF_V is not resulting in an error because ΔF_V varies linearly with ε and its influence has already automatically been taken into account when calibrating the preloaded sensor. The limit of measuring F_s is again reached when $F_s > F_V$ and the sensor becomes unloaded and lose. Then the bolt would rapidly start to expand because of its low spring constant until it fractures. Because preloading bolts are usually stressed to close their elastic limit, care must be taken that F_s will never come close to F_V for safety reasons! A good rule of thumb is to insure that $F_s < 0{,}7F_V$.

A further advantage of such preloaded force sensors is the particularly good linearity of typically $<\pm 0{,}1\,\%$ FSO and minimum hysteresis, because the micro gaps between all contact surfaces are closed to a maximum degree by the preload. This increases overall rigidity which in turn means highest possible natural frequency.

Short and thick preloading bolts could be used – then, ordinary steel would be sufficient – but there are a number of disadvantages. As Fig. 6.7c shows, k_V and k_A are then of the same order of magnitude. This means the force shunt is now about 50% and the resulting sensitivity of the sensor is halved. The preloading force $F_V + \Delta F_V$ varies substantially over the measuring deformation ε of the sensor and can easily exceed the value corresponding to the elastic limit of its material long before $F_s = F_V$. If such an unfavorable design can not be avoided, all aspects must be considered and a complete worst-case stress analysis should be made.

Load-washer-type force sensors which are already mounted and preloaded between 2 steel nuts are known as force links (Fig. 6.8), i.e. force sensors for

Figure 6.8 Force measuring elements ("force links") for compression and tension force with calibrated ranges from ±2,5 (with M5 connecting thread) to 120kN (with M30 connecting thread) (Courtesy of Kistler)

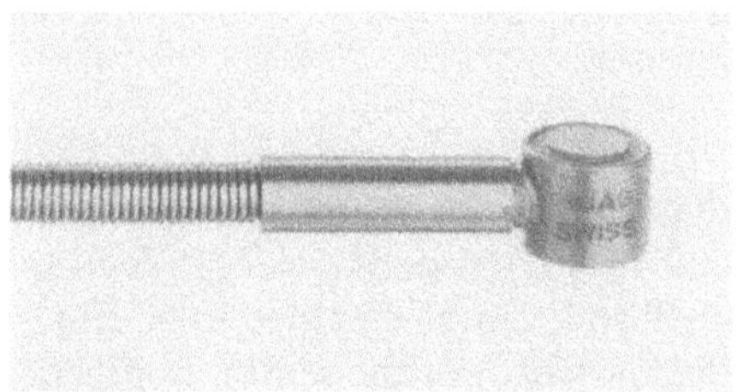

Figure 6.9 Miniature force sensor for 2,5kN with an outside diameter of only 6mm (Courtesy of Kistler)

compression and tension forces, already calibrated and ready to use. They are available with ranges from less than ±1 kN to over ±100kN.

Load washers and force links have, thanks to their high rigidity, a very wide frequency range. Depending on the mass and the elastic properties of the measuring setup, natural frequencies of several kHz and up to far beyond 10kHz are possible.

In miniature force sensors (Fig. 6.9) the quartz plates are directly preloaded by the walls of their housing (see also Fig. 5.7). Here the preload is only needed to hold all parts in place and to minimize the micro gaps in the contact surfaces. Since such sensors measure only compression forces, they do not need mounting threads.

All force sensors described so far use transduction elements exploiting the longitudinal piezoelectric effect in quartz. Usually in pairs, these elements are mechanically in series and electrically in parallel. Their nominal sensitivity is always around 4pC/N or a little less, depending on the force shunt by the mounting. Modern charge amplifiers have a noise level of around 10fC which corresponds with that sensor sensitivity to about 10fC/(4pC/N)=2,5mN. For practical purposes triple the value corresponding to the noise level is usually taken as clearly detectable smallest change in the measurand, i.e. the threshold of all these sensors – irrespective of their maximum range – is about 10mN.

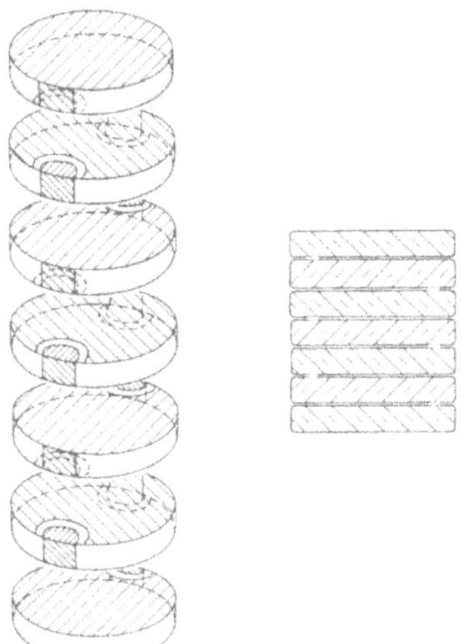

Figure 6.10 Micromodule with 7 quartz plates for the longitudinal effect (after AVL)

Basically this threshold can be somewhat lowered by using sophisticated signal processing techniques which are rarely realistic in a common measurement situation. However, there are 3 solutions for building sensors with a higher sensitivity hence a lower threshold: 1) use a large number of quartz plates cut for the longitudinal effect, 2) use quartz elements cut for the transverse effect and optimize their ration of thickness to length, and 3) use a piezoelectric material with a higher sensitivity such as piezoelectric ceramics.

An example for solution 1 is the so-called micro module shown in Fig. 6.10. Quartz plates cut for the longitudinal effect are first fitted with vacuum-deposited electrodes laid out to make appropriate contact for electrically paralleling their outputs when stacked. A number of such plates is then stacked and heated to solder them together via their electrodes. Soldering also eliminates micro gaps which would reduce the rigidity of the module. Yet the gain in sensitivity is small because the nominal sensitivity of 2 pC/N per plate is only multiplied by the number of plates used. The module shown has a sensitivity of about 14 pC/N.

A much higher sensitivity can be obtained with solution 2, using the transverse effect in quartz. The sensors illustrated in Fig. 6.11 have a column-type quartz element whose long axis corresponds to the crystallographic y-axis. The sensitivity of such an element grows proportionally with its length and inversely proportional with its cross section (more precisely: with its thickness in the crystallographic x-axis). With a sufficiently slender element a sensitivity of about 50 pC/N can be obtained which is 25 times more than that of a single x-cut quartz plate.

While very slender – long and thin – elements would have even higher sensitivities, there is a practical limit. The maximum load decreases rapidly because no longer the normal stress in the quartz is the limit but the critical load for buckling of the element. Another limit is set by the different coefficients of

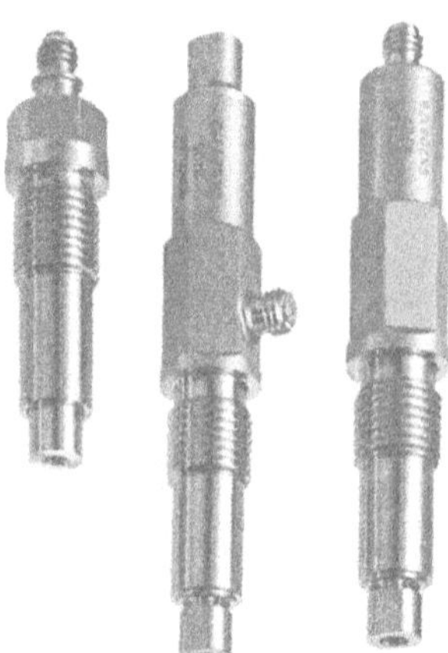

Figure 6.11 High-sensitivity force sensors with quartz elements for the transverse effect (Courtesy of Kistler)

thermal expansion in quartz and steel which result in a high temperature error that can not be entirely compensated for.

The third solution is to use piezoelectric materials with a high sensitivity such as piezoelectric ceramics. They allow to achieve sensitivities of around 50 pC/N, too. Force sensors with such elements have several drawbacks. Piezoelectric ceramics have a linearity of only about ±3 %FSO while that of quartz is typically below ±0,3 %FSO. All ceramics inherently have hysteresis while quartz is practically free of hysteresis. Also the thermal properties of sensors with ceramic element are less favorable than those of quartz sensors, in particular the strong pyroelectric effect exhibited by ceramics. The insulation resistance of ceramics is markedly lower that that of quartz which makes quasistatic calibration as well as quasistatic measuring impossible. These aspects are probably the reason why there are practically no piezoelectric force sensors with ceramic transduction elements on the market.

6.3 Multicomponent Force Sensors

Multicomponent force sensors are mainly built with quartz and for high-temperature or high-sensitive applications also with crystals of the CGG-group (which have the same matrix of piezoelectric coefficients as quartz) as transduction elements. The design is basically the same as in single-component force sensors, but with a higher number of quartz elements.

3-component force sensors have a pair of quartz elements cut for the longitudinal piezoelectric effect for measuring the compression force (the z component F_z) and a pair of quartz elements cut for the shear effect each for measuring the shear components (the x component F_x and the y component F_y) of the acting force (Fig. 6.12). Shear forces are transmitted by friction which means

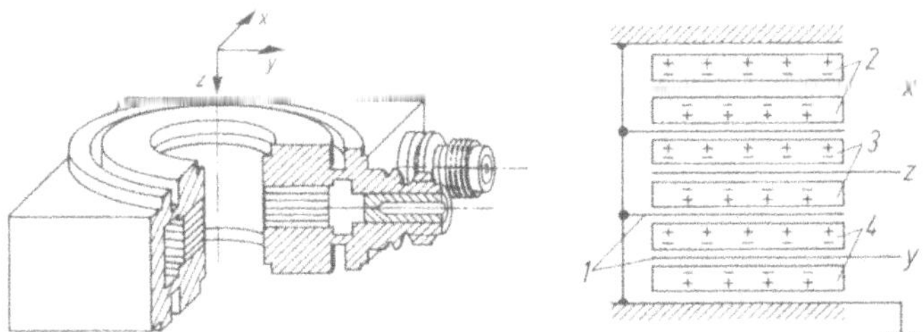

Figure 6.12 Design of a 3-component force sensor (after Kistler). *1* electrodes, *2* quartz plates for the shear effect, measuring the *x*-component of the force, *3* quartz plates for the longitudinal effect, measuring the *z*-component of the force, *4* quartz plates for the shear effect, measuring the *y*-component of the force

Figure 6.13 3-component force sensor (24·24·10mm). Ranges for F_x and F_y are ±2,5kN, for F_z ±510kN (with standard preload of 25kN) (Courtesy of Kistler)

that such a sensor must be mounted under high preload (see 6.6). Usually a coefficient of friction of $\mu=0{,}1$ is used to calculate the admissible shear force. In some designs using special materials, a coefficient of even $\mu=0{,}15$ can be applied. Fig. 6.13 shows a 3-component force sensor which measures the 3 orthogonal components of any force acting on it. An extremely miniaturized quartz 3-component sensor was designed to measure the load exerted by dental prosthesis on implanted anchorage posts (Figs. 6.14 and 6.15) [Piotti and Sirtes 1996].

Generally multicomponent force sensors are used in groups of 3 or 4, built into so-called dynamometers or force plates. Piezoelectric sensors have the particular property that several sensors (selected to have the same sensitivity) can be connected electrically in parallel directly to a single charge amplifier whose output then corresponds to the algebraic sum of all measurands (e.g. forces) on all connected sensors (see 11.6). The 3-component dynamometer in Fig. 6.16 is built with 4 sensors as shown in Fig. 6.13 and is mainly intended for measuring the cutting forces in machining. For turning, the cutting tool is fixed on the dynamometer while for milling and grinding, the work is attached to the dynamometer. All x-, y- and z- outputs are paralleled respectively and led to the 3 output connections of the dynamometer. The four 3-component sensors are selected with their respective sensitivities to be within narrow tolerances so that the dynamometer is now functioning like a single 3-component force sensor. The 4 sensors will not be loaded equally, depending on the point of force application, yet the dynamometer output is independent of the point of force application. A torque acting on the dynamometer will load the sensors, but does not influence the

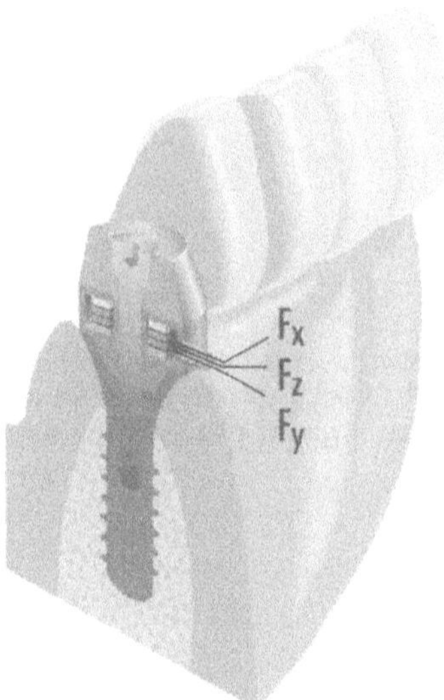

Figure 6.14 Miniature 3-component force sensor built into a dental implant anchorage post (Courtesy of Kistler)

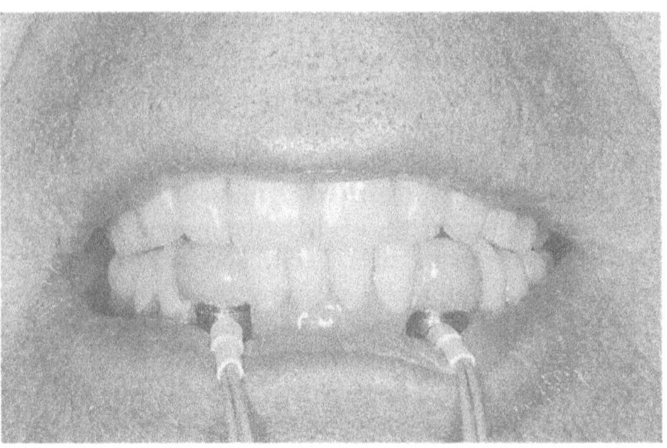

Figure 6.15 In-vivo 3-component force measurement of load on a dental prosthetics with 2 sensors as shown in Fig. 6.14 (Courtesy of Kistler)

output of the dynamometer because a torque has – by definition – no external resultant force (see 6.1).

However if the outputs of the 4 sensors are not paralleled but processed individually, not only the 3 orthogonal force components can be determined but also the 3 components M_x, M_y and M_z of the resulting moment vector about the origin of the coordinate system defined by the 4 sensors. This method is described in 6.6.

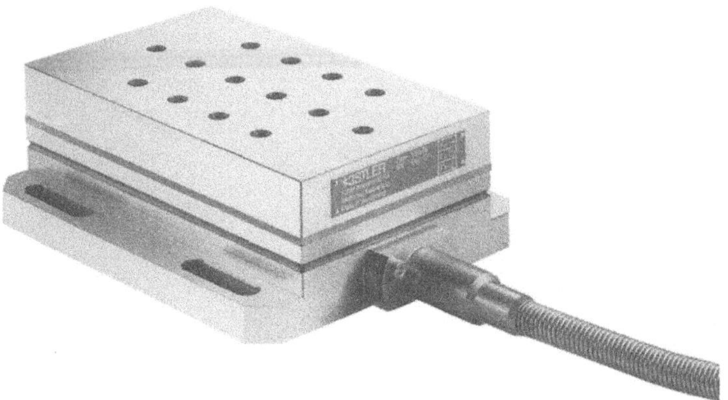

Figure 6.16 3-component dynamometer. Ranges for F_x and F_y are ±5 kN, for F_z −5...10 kN. Natural frequency: ≈3 kHz (Courtesy of Kistler)

6.4 Torque Sensors

The arrangement of shear-sensitive quartz elements shown in Fig. 6.17 can be used to measure torque. The quartz elements are aligned with their sensitive axis tangent to the circle. They must be mounted under the required mechanical preload because the shear forces are transmitted by friction. All quartz elements are electrically in parallel so that their total output is proportional to the acting torque. With a sensor having such a transduction element a dynamometer as shown in Fig. 6.18 can be built which e.g. allows measuring torque and feed force in drilling and tapping.

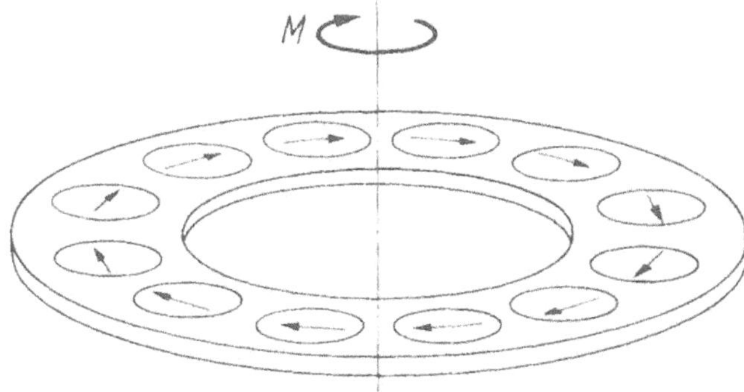

Figure 6.17 Arrangement of quartz disks cut for the shear effect for measuring a torque. The individual quartz disks are held in the correct position by a highly insulating potting compound (after Kistler)

Figure 6.18 4-component dynamometer. Ranges for F_x and F_y are ±5 kN, for F_z −5...200 kN, and for M_z ±200 Nm (measured with an arrangement as shown in Fig. 6.17). Natural frequency: >4 kHz (Courtesy of Kistler)

Although it would be possible to use a transduction element (made of quartz or another piezoelectric material) that responds directly to torque (Fig. 5.3f) such designs have not been successful in practice. One of the reasons is that the entire torque would have to be transmitted by the transduction element which would limit the possible range. Also it is very difficult to introduce the torque mechanically to the transduction element without undesired strain, causing poor linearity and uneven loading of the element, further limiting the range.

An entirely different method for measuring torque is offered by the multicomponent measuring systems based on 3 or 4 3-component force sensors (6.6).

6.5 Cross Talk in Single- and Multicomponent Force Sensors

Ideal force sensors give only an output when they are loaded by a force acting in their sensitivity axis. A force normal to that axis should not produce any output. However, real force sensors give an output also to a force normal to their sensitive axis. This phenomenon is called cross talk (see 4.2.6).

In practice the significance of cross talk is often overlooked or just ignored although it can have a significant influence on the accuracy of the measuring result. Fig. 6.19 gives an illustrative example. The load-bearing top plate is mounted via 2 single-component force sensors on a solid base plate. The top plate is not infinitely rigid and will bend when loaded by the force F_z to be measured. This bending produces the shear forces F_{x1} and F_{x2}, which subject the sensors to a shear load. The vector sum of these 2 shear forces is zero because they are internal forces of a statically overdetermined system. The also acting moments are disregarded here for simplicity. If the 2 sensors were ideal, the forces F_{x1} and F_{x2}

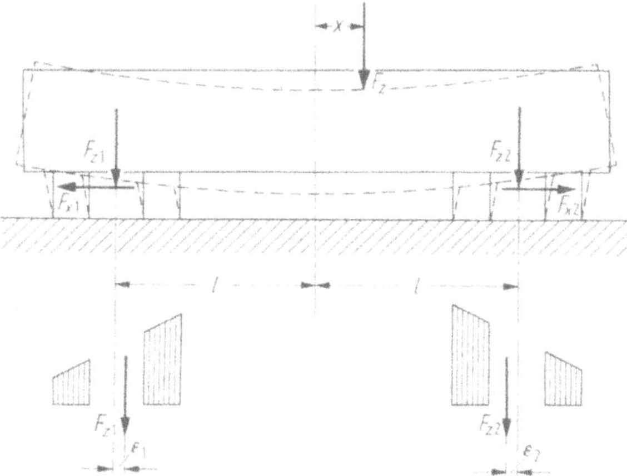

Figure 6.19 Cross talk due to internal forces and measuring error caused by deformation

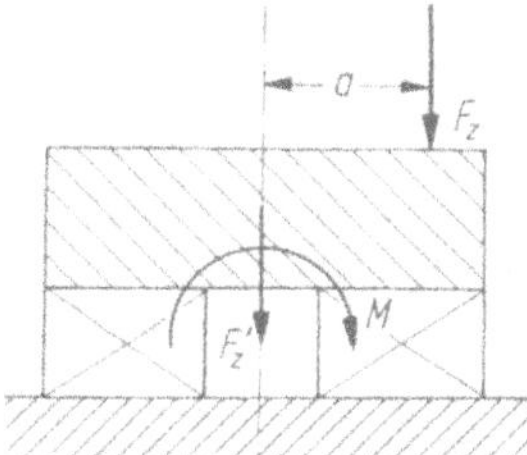

Figure 6.20 Cross talk due to eccentric loading

acting normal to the sensitivity axis of the sensors would not produce any output in them.

In reality there is some cross talk. Assuming a cross talk value of 2% this means that the measured value for is falsified by 2% of F_{x1} and F_{x2} respectively. The error so caused can be substantial, besides that F_{x1} and F_{x2} are not a linear function of F_z.

Another example, shown in Fig. 6.20, is a force sensor which is eccentrically loaded by the force F_z. The eccentric force F_z can be replaced by the centric force F'_z and the moment $M=F_z \cdot a$ about the center line of the loaded surface of the sensor. The sensor is loaded by F'_z and M. While F_z is measured correctly, a cross talk of M into the sensitive axis of the sensor (for F_z) will falsify the result. Note that here the cross talk can not be expressed in percent but in N/Nm because F_z and M have different units. Only if F_z acts in the center of the sensor will the cross talk of M into F_z have no influence on the result because $M=0$.

Even more critical is cross talk in multicomponent force sensors. If a force F is acting exactly in the direction of the z axis on the sensor, the x- and y-outputs of a perfect sensor would be zero. A real sensor gives some output on its x- and y-axes, too, because of the cross talk. When the direction of the measured force is

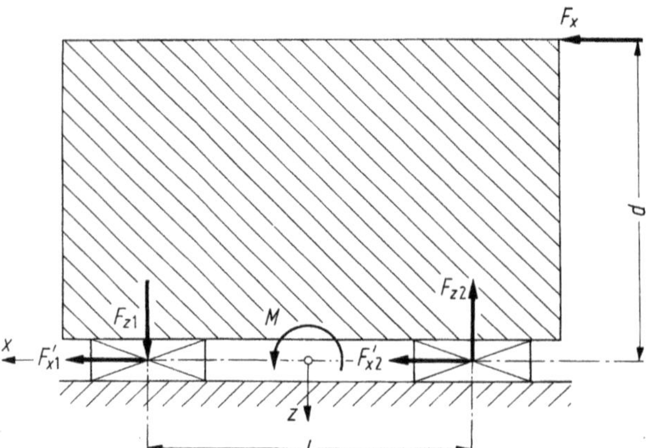

Figure 6.21 Cross talk due to different sensitivities of mounted sensors

calculated from the 3 force component measured, the result will not be correct. The cross talk not only can produce an error in the value of a measured force, but also in the direction of the force as determined from the sensor outputs.

Piezoelectric 3-component force sensors with quartz elements offer very low cross talk values, typically below 1%, provided they are of correct design. A prime condition is the precise working of the quartz elements (see 5.5.1). The mechanically loaded surfaces of the quartz plates must be normal to the respective crystallographic axes within very close tolerances (an error in angle of 35' results in 1% cross talk!). In addition, several criteria must be observed when mounting such sensors (see 6.6).

In systems based on several sensors mounted between a base and a top plate cross talk can have still other causes. In Fig. 6.21 the massive top plate is mounted on 2 preloaded force sensors. The 2 sensors are single-component sensors for the z-direction and the system is to be used to measure only the z-component of forces acting on the top plate. With ideal sensors, the system should give no output for a force *Fx* acting normal to the z-direction. The force F_x can be replaced by the force F'_x and the moment M whereby F_x acts in the centerline of the F_z-sensitive quartz plates of the sensors. The force F'_x splits into F'_{x1} and F'_{x2}.

A first type of cross talk is the inherent cross talk from the x- into the z-direction of the sensors due to imperfect orientation of their F_z-sensitive quartz plates which suggests a nonexistent force component in the z-direction. A second type of cross talk originates from small differences of the z-sensitivities of the the 2 sensors. The moment $M = d\,F_x$, equivalent to the force couple F_{z1} and F_{z2}, would not lead to an output if the z-sensitivities of both sensors were identical because $F_{z1}+F_{z2}=0$. In reality, sensitivities vary slightly between sensors so a cross talk of M into F_z results. This cross talk can be caused also by only a torque acting on the top plate, i.e. without a force F_x acting. Such type of cross talk can only be minimized by carefully selecting sensors of the same sensitivity in the respective axes and preloading elements of exactly the same dimensions, i.e. machined to within very close tolerances. Still another source could be poor mounting of the sensors, such as preloading bolts with their axes not exactly normal to the mounting surfaces.

Finally there is a possibility of electrical cross talk in multicomponent force sensors. If e.g. a 3-component force sensor is used with only one of its outputs connected to a charge amplifier (the other outputs being left open because they are not needed) electrostatic voltage $U=Q/C_{sensor}$ can build up on the non-connected outputs which may reach several hundred volts. There is usually no electrical shielding between the 3 channels within such sensors for design reasons. As long as all outputs are connected to a charge amplifier no voltage can develop across any of the outputs and no electric cross talk occurs. A electrostatic voltage on a non-connected output however will electrically influence the signal in the connected measuring channel. Therefore it is imperative in such cases that the outputs of the channels not used be short-circuited by placing so-called short-circuit caps on the respective connectors of the sensor to prevent electrostatic voltage from building up. In the channel used the voltage remains practically zero anyhow because of the connected charge amplifier.

6.6 Mounting Force Sensors

The extremely high rigidity is the major advantage of piezoelectric force sensors. Their measuring deflection at full load is typically less than about 20 μm which not only minimizes errors in static measurements (Figs. 5.1 and 5.2) but is also a sine qua non for a high natural frequency – hence a wide measuring frequency range – of the measuring setup. Therefore special attention must be paid to the mechanical conditions when mounting such sensors.

This is best illustrated by the example of measuring the clamping force under a bolt head. While one might generally consider the head of a bolt as quite a rigid structure, it must be remembered that any structure, no matter how thick or massive, will deform when loaded. A bolt head always deforms when the bolt is tightened (Fig. 6.22a). If a load washer is mounted directly under the bolt head as

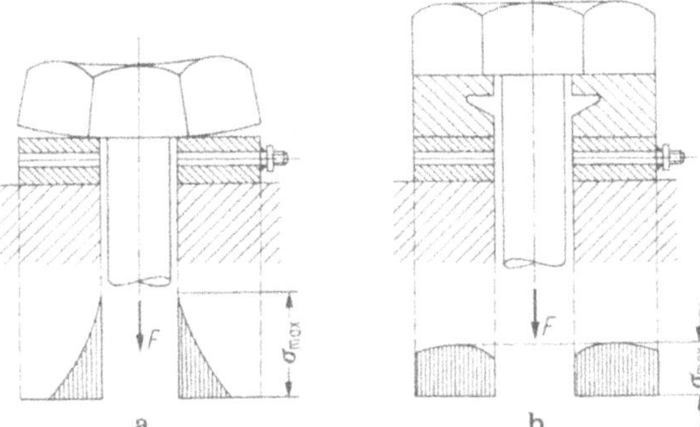

Figure 6.22 Measuring the tension force in a bolt with a load washer

sensor for measuring the clamping force of the bolt, the resulting stress distribution across the piezoelectric transduction elements (i.e. the quartz plates) in the sensor will not be homogenous but show a very high peak value at its inner edge. *As long as the peak stress does not exceed the allowable stress for quartz, the force will be measured correctly because the electric charge yielded by the quartz plate is the integral of the stress across the loaded surface* [Kenner 1975]. In practice a generally accepted conservative limit for quartz is $\sigma_{max} \approx 150\,N/mm^2$. If this value is considerably exceeded, the quartz may twin (see 3.2.4) or even break, i.e. the sensor will be destroyed. Therefore the deformations of the parts between which the sensor is mounted and the resulting non-homogenous stress distribution must be carefully considered in the design of the sensor mounting.

Such deformations are in order of a few µm and therefore difficult to imagine. A good way to visualize deformations is to imagine the structure and the sensor to be made of a very elastic material such as rubber or plastic foam – or even to make a model using such materials. The critical points which could lead to stress peaks then become highly visible and allow to make the necessary adjustments in the design of the sensor mounting.

In the example of the bolt head this can easily be achieved by inserting a thick steel washer with an undercut in its bore (Fig. 6.22b). The peak stress at the inner edge due to the deformation of the bolt head is no longer directly transmitted to the quartz element because the undercut leads to a more symmetrical and homogeneous stress distribution. Most important is that the maximum stress is no longer at the edge of the quartz plate, but somewhere inside the loaded surface. Edge loading is most critical for causing the quartz plate to crack. An easy-to-remember rule is: *Install a piezoelectric force sensor as if it were made of glass.* Glass can take a very high loading provided the stress distribution is as homogeneous as possible and that there is no stress peak at the edge (edge loading) or at any point within the loaded surface, i.e. a point loading (e.g. caused by a speck of dust or a grain of sand).

In this context a remark regarding the overload specified for such force sensors (load washers and similar designs) must be made. In practice it is rarely possible to make a detailed stress analysis and therefore it is advisable to use a conservative value for overload. Not all manufacturers of such sensors handle this in the same way which means that for the same size of a sensor sometimes considerably different overload values are specified. Unless a stress analysis or practical evaluation tests can be made it is recommended to multiply the limit stress of $150\,N/mm^2$ by the load bearing area of the quartz, which is usually the same as the area of the load bearing face of the sensor.

Manufacturers of similar types of sensors do not always use the same values for determining the range of the sensor which may be confusing for the user. If e.g. the range specified for a certain size of load washer by manufacturer A is much higher than that specified by manufacturer B it simply means that manufacturer A leaves the user of the sensor with a much smaller safety margin! Therefore, sensors that are used in experimental situations should be chosen with a higher safety margin while sensors that are installed e.g. in process control and monitoring systems with

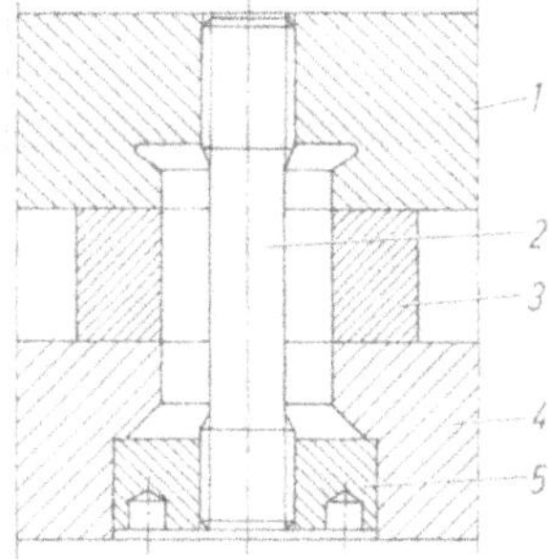

Figure 6.23 Standard mounting of a 3-component force sensor (after Kistler). *1* top plate, *2* preloading bolt, *3* sensor, *4* base plate, *5* preloading nut

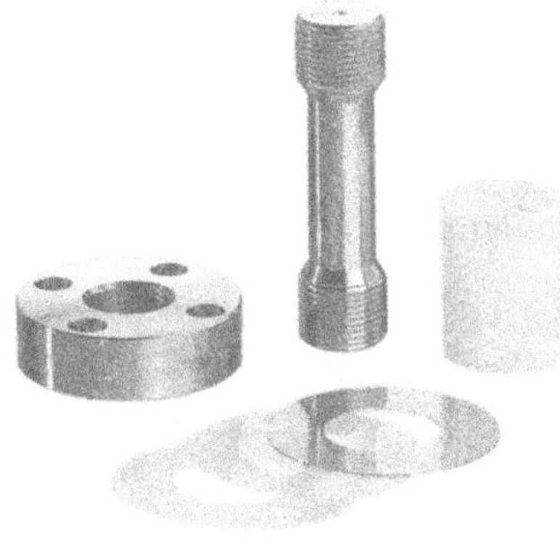

Figure 6.24 Preloading elements for mounting shown in Fig. 6.23: high-strength bolt, special nut, sliding washer (beryllium) and centering sleeve (Delrin®) (Courtesy of Kistler)

well defined, repeatable mounting conditions can be selected with a smaller safety margin. If the sensor is really loaded by a very even stress distribution (like "hydrostatic" loading) , then it can easily support a total force corresponding to a stress level of $300\,N/mm^2$ or more across the quartz plates. This is valid only at temperatures up to about 200 °C, because at higher temperatures, twinning may occur in the quartz element (see 3.2.4).

In general, overload should be set to at least 1,3 to 1,5 times the span (specified range) of a force sensor to minimize the danger of changing the sensor's specifications beyond tolerance by exceeding the upper limit of he range accidentally.

Another example is the standard mounting of 3-component force sensors shown in Fig. 6.23. They must always be mounted under high mechanical preload because shear forces are transmitted by friction. A sufficiently even stress distribution across the load-bearing sensor surface can be achieved by design details such as reduced contact area, centered on the sensor face and undercuts. Special high-stregnth preloading bolts, nuts and accessory parts (Fig. 6.24) are offered by manufacturers of 3-component force sensors to facilitate correct mounting.

3-component force sensors are usually installed in groups of 2, 3 or 4 sensors to form so-called dynamometers and force plates whose design has to aim at even load distribution on the sensors. The base plates and top plates of such dynamometers and force plates have to be sufficiently rigid to keep their deformation within limits especially under point-type loading. The base plate is often fixed on another base such as a machine tool table or a massive steel plate grouted in a heavy concrete foundation which will add to the rigidity. Then the base plate may even be quite thin and elastic provided it is always attached to such a solid base. The same may apply to the top plate when e.g. solid parts are attached to it. This stiffening effect works only if the contact surfaces are fine-machined and absolutely flat. Also there must be a sufficient number of bolts holding the parts together.

In summary the outstanding properties of piezoelectric force sensors can only be fully exploited and good measuring results obtained when the mounting is done in a correct way.

6.7 Measuring 3 Forces and 3 Moments (6 Components)

The piezoelectric 3-component force sensors and the multicomponent measuring systems based on them have opened up vast new fields of applications. Often they allowed for the first time to accomplish a great variety of measuring tasks hitherto considered difficult or even impossible to solve as is illustrated by the examples following.

Wheel force dynamometers for measuring tire loading on a rolling road test stand (Fig. 6.25) are used by the tire industry to develop and test tires, i.e. to

Figure 6.25 Multicomponent dynamometer, used as wheel force dynamometer (Courtesy of Kistler). **a** before assembly (note that here, in each corner, a 1-component and a 2-component sensor are combined), **b** after preloading and complete assembly (the wheel bearings will be inserted into the center bore)

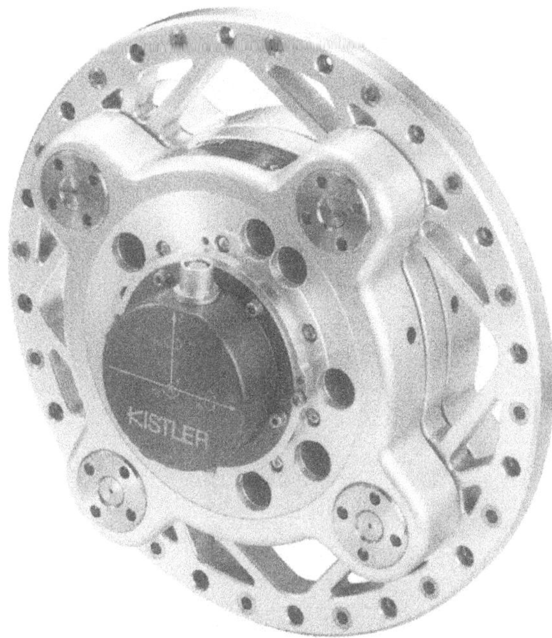

Figure 6.26 Rotating wheel dynamometer (RWD), made of aluminum alloy. The RWD attaches with an adapter to the standard hub of the car while a standard BBS light alloy rim is attached to the outer hole circle. In each corner, 2 3-component sensors are arranged in a push-pull fashion, sandwiching the center plate supporting the rim between the outer plates of the RWD, fixed to the hub. The outputs of the charge amplifiers are transmitted by sliprings to the nonrotating part with the electrical connector on top. An angle encoder provides the input for continuous coordinate transformation, i.e. for transforming the signals of the rotating sensors into force signals referring to a nonrotating coordinate system (Courtesy of Kistler)

determine the tire characteristics. In automotive engineering, rotating wheel dynamometers (Figs. 6.26 and 6.27) allow measuring the wheel load while driving and to study a car's drivability as well as e.g. the characteristics of the suspension. Such dynamometers measure reliably at speeds up to over 240km/h [Berther and Burkard 1999; Burkard and Berther 2000; Burkard and Calame 1998; Mandel and Gautschi 1997/8; Martini 1983].

Cutting force dynamometers [Fischer 1968] can measure cutting forces in all types of metal working (Figs. 6.16, 6.18 and 6.28). Thanks to the very low threshold and the high rigidity of quartz sensors, small forces in the order of a few mN occurring e.g. in micromachining and wafer slicing can be captured reliably [Gautschi 1972; Kirchheim et al 1996; Spur et al 1993; Stirnimann and Kirchheim 1997]. Using 3-component force sensors with transduction elements made of a crystal from the CGG group (see 3.5), very compact dynamometers with a

Figure 6.27 Rotating wheel dynamometer (see Fig. 6.26) fully installed on a test vehicle (Courtesy of Kistler)

threshold in the range of mN, yet measuring ranges of ±250 N and a high natural freqeuancy of over 5 kHz can be built [Gossweiler and Cavalloni 1999].

Force plates (Fig. 6.29) serve to measure ground reaction forces (magnitude, direction and location, i.e. location of the point of application on the ground) in biomechancis. In biomechanics of sports, the forces exerted by the feet of athletes in walking, running or jumping can be analyzed. When measuring ground reaction forces in weight lifting, the force plate can be designed with a large enough range to take into account force peaks (can be several tons) occurring when the athlete accidentally drops his weight, yet – thanks to the inherent very low threshold of piezoelectric sensors – the desired forces can still be measured without any compromise in accuracy. Similarly, ground reaction forces can be

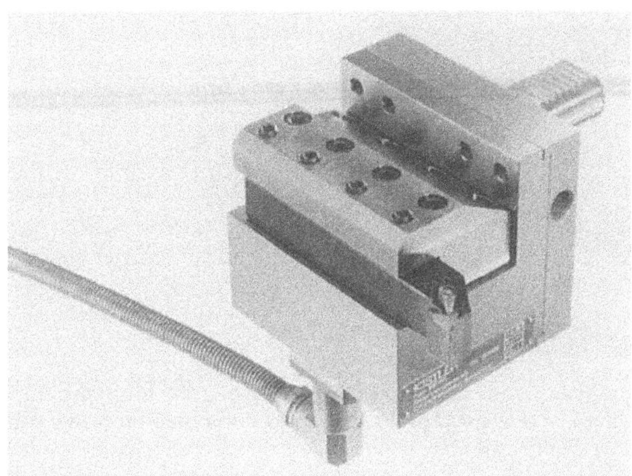

Figure 6.28 3-component toolholder dynamometer (installs directly in the turret of a NC-machine) for measuring cutting forces in turning (Courtesy of Kistler)

Figure 6.29 Multicomponent force plate (Courtesy of Kistler). **a** force plate for applications in biomechanics and automotive engineering, **b** base frame with the 4 preloaded 3-component force sensors (top plate removed)

The shoe industry uses force plates to determine damping and shock absorbing characteristics of both, sports and regular shoes. Safety aspects such as slipping can be studied too, because the ratio of horizontal to vertical force is the coefficient of friction which thus can be measured dynamically and continuously, also under varying vertical load. Thanks to the very compact design of quartz 3-component

Figure 6.30 Force plates with top plates made of safety glass for viewing the footprint from below (each plate measures 0,5 · 1 m). Note the compact size of the 3-component force sensors in the corners of each plate (Courtesy of Kistler)

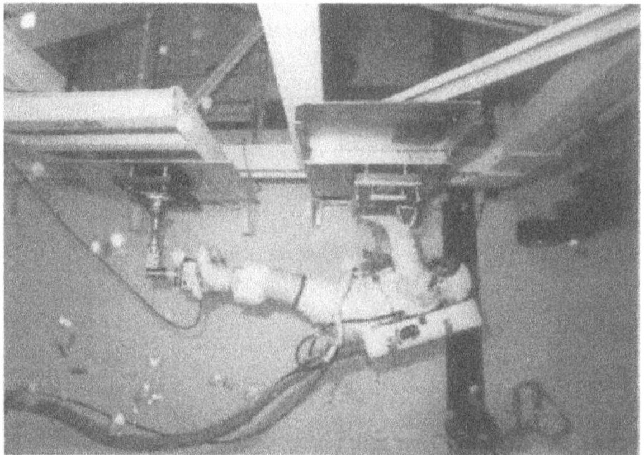

Figure 6.31 Force plates with quartz multicomponent force sensors in a water tank for ergonomic measurements on astronauts during practice for working in space. Similar force plates and dynamometers are also used for experiments during space missions (Courtesy of Kistler/NASA)

sensors, the top plate can also be made of glass (usually multilayer safety glass, see Fig. 6.30) and film or video recordings of the footprint (athlete's or patient's foot, or footprint of a tire) can be made and, in real time, correlated and combined with the force magnitude, direction and point of application measured by the force plate. Ergonomics is another field of application, be it in industry or e.g. in astronautics (Fig. 6.31).

For crash testing of cars, crash walls are equipped with 3-component force plates attached to the impact face for recording the impact force which can be over 100 tons (Fig. 6.32). Usually, segments mounted on 4 to 8 sensors each are assembled on a heavy steel plate, attached to a concrete block of several hundred tons, and machined to be flat within very close tolerances where the sensors are

Figure 6.32 Dynamometer system for crash testing of cars. Smallest plates are 250·250 mm, each mounted on 4 3-component force sensors. In the picture, the impact face is turned 30° out of a position normal to the impact direction to study so-called "glancing impacts". The impact forces are resolved into a component normal to the face and parallel with the impact face (Courtesy of Kistler)

mounted. Natural frequencies of over 1 kHz can be achieved which allows to analyze the dynamics of the impact in detail [Berther and Jacobson 2000].

Other unusual applications include e.g. measuring the 3 force components in the contact point of an electric locomotive's pantograph with the trolley wire (overhead contact wire) – a problem present especially in high-speed trains – or of the forces occurring in linear motors and magnetically suspended trains. Here the complete insensitivity of quartz transduction elements to electric and magnetic fields can be fully taken advantage of. However it may be necessary to make the sensor housing of an antimagnetic material (e.g. austenitic steel) when strong electric fields are present because error signals can be produced when e.g. in a load washer the 2 rings sandwiching the quartz transduction elements will attract each other due to induced magnetic fields and give an output which is only caused by the external electric field. An example for this is measuring the clamping force in spot welding where a high current flows through the electrode that passes through the center bore of the sensor.

Still another example is measuring the braking force and the coefficient of friction. The special dynamometer shown in Fig. 6.33 has a pair of 2-component sensors and was used to measure the braking and the clamping force (allowing to determine the coefficient of friction) in an experimental setup to study the characteristics of the emergency brakes used on aerial cable cars. During simulation of an emergency stop, the coefficient of friction can be determined dynamically and continuously as a function of the clamping force. Various

Figure 6.33 Experimental brake dynamometer for dynamically measuring the braking force and the coefficient of friction in the emergency brake of an aerial cable car. Note the 2 cables, corresponding to the brake clamping force and the friction force (Courtesy of Kistler/ETHZ)

materials for the brake pads and their fading characteristics could be studied in situ. Although temperature on the brake pads reached over 600 °C (setting the grease on the cables on fire) the quartz sensors measured with minimum thermal error [Gassmann 1979].

Multicomponent measuring systems are generally built by using a group of 3 or 4 3-component sensors. The general principle of 6-component measuring systems is described for an arrangement with 4 sensors. The corresponding formulae for systems with 3 sensors can be derived analogously and the measuring possibilities are exactly the same. In general, it is only possible to determine the 3 orthogonal components of the resulting force acting on the system and the 3 components of the resulting moment vector about the origin of the Cartesian coordinate system defined by the sensors. While the direction of the resultant force vector can be determined, it is not possible to determine the position in space of the resultant force vector, as it is shown by the following analysis.

Consider 4 identical 3-component force sensors sandwiched between 2 infinitely rigid steel plates under preload (Fig. 6.34). In the general case, a force F and a torque T can act on the top plate. The force F can be resolved into its components F_x, F_y and F_z, the torque into T_x, T_y and T_z.

The 4 sensors constitute a measuring system that measures in reference to the Cartesian coordinate system (x, y, z) defined as follows: the xy-plane is defined as the central plane of the quartz plates sensitive in the z-direction and the origin is in the center of the rectangle defined by the 4 sensors.

While the components F_x, F_y and F_z are measured independently of the point of application of F, the moments measured by the system are relative to the 3 coordinate axes. The torque vector $\boldsymbol{T}$ can be shifted to go through the origin of the coordinate system without changing its influence on the measuring system

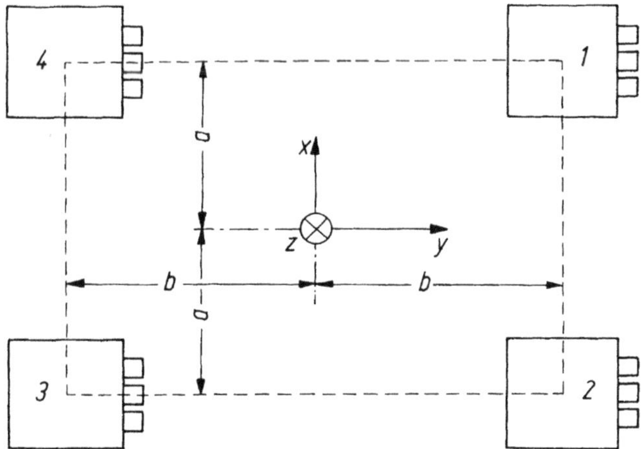

Figure 6.34 Multicomponent measuring system with four 3-component force sensors

(property of a torque). In the general case the action line of the force vector ***F*** does not go through the origin of the coordinate system and produces a moment M_F about the origin which is measured by the system in addition to the acting torque T. Therefore the moment M referred to hereafter is always understood to be the vector sum of the acting torque T and the the moment M_F produced by F about the origin.

By calling the force components measured by the force sensors $F_{x1}, F_{x2} \ldots, F_{y1}, F_{y2} \ldots$ and $F_{z1}, F_{z2} \ldots$, we can write the following equations for the components of the acting force F:

$$\begin{aligned} F_x &= F_{x1} + F_{x2} + F_{x3} + F_{x4}, \\ F_y &= F_{y1} + F_{y2} + F_{y3} + F_{y4}, \\ F_z &= F_{z1} + F_{z2} + F_{z3} + F_{z4}. \end{aligned} \tag{6.1}$$

Designating the distance between the sensors in the x-direction with $2a$ and in the y-direction with $2b$, and by calculating the moments about the 3 coordinate axes, we get:

$$\begin{aligned} M_x &= b(F_{z1} + F_{z2} - F_{z3} - F_{z4}), \\ M_y &= a(-F_{z1} + F_{z2} + F_{z3} - F_{z4}), \\ M_z &= b(-F_{x1} - F_{x2} + F_{x3} + F_{x4}) + a(F_{y1} - F_{y2} - F_{y3} + F_{y4}). \end{aligned} \tag{6.2}$$

Therefore, after converting each of the altogether 12 output signals of the 4 sensors into proportional voltages by means of charge amplifiers and then processing them according to the above equations, all 6 components which are possible to get from the mechanical analysis can be obtained, i.e. F_x, F_y, F_z, M_x, M_y and M_z. The magnitude of the resulting force vector is

$$F = \sqrt{F_x^2 + F_y^2 + F_z^2} \tag{6.3}$$

and the direction of the vector is defined by the directional cosines

$$\begin{aligned} \cos\alpha &= \frac{F_x}{F}, \\ \cos\beta &= \frac{F_y}{F}, \\ \cos\gamma &= \frac{F_z}{F}. \end{aligned} \tag{6.4}$$

However, the position of the force vector $\boldsymbol{F}$ in space can not be determined – only its direction is known!

Therefore, a multicomponent force and moment measuring system provides only the 3 components of the resulting force vector, its direction (not position!) and the 3 components of the resulting moment vector about the origin of the coordinate system. This can also be shown by analyzing the external forces and torques acting on the measuring system. Consider the system shown in Fig. 6.35.

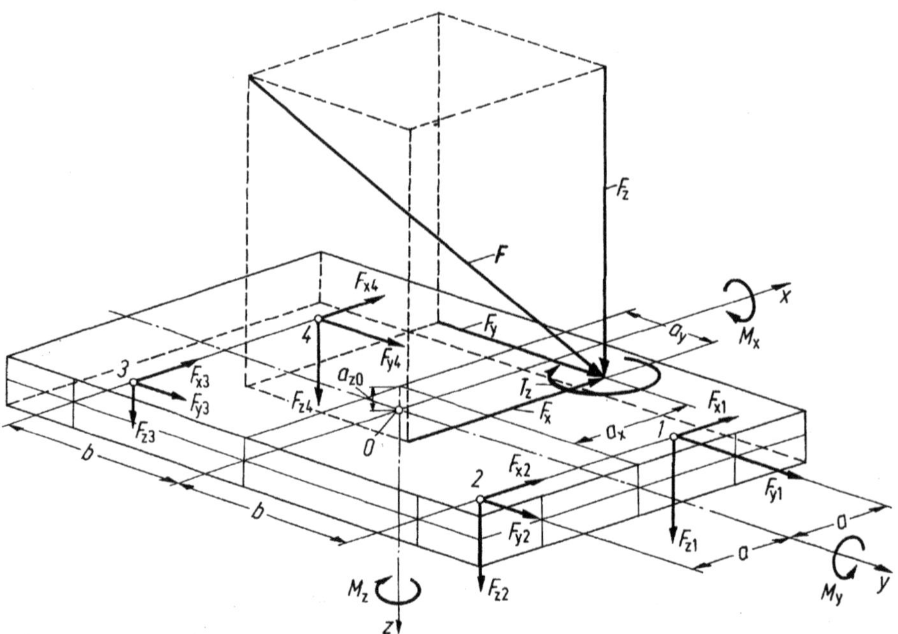

Figure 6.35 Multicomponent measuring system referring to the force plate shown in Fig. 6.29. Note: The vertical offset a_{z0} shown in the picture has a negative value because the positive z-axis points downward

At first, only the force vector $\boldsymbol{F}$ is acting on the system. $\boldsymbol{F}$ can be replaced by its components F_x, F_y and F_z for the purpose of calculation. The components of the resulting moment vector about the origin are calculated as follows:

$$
\begin{aligned}
M_x &= a_y F_z - a_z F_y, \\
M_y &= -a_x F_z + a_z F_x, \\
M_z &= a_x F_y - a_y F_x.
\end{aligned} \tag{6.5}
$$

Because the system measures F_x, F_y, F_z, M_x, M_y and M_z, for obtaining a_x, a_y and a_z, these equations can be rewritten as

$$
\begin{aligned}
0\ a_x + F_z a_y - F_y a_z &= M_x, \\
-F_z a_x + 0\ a_y + F_x a_z &= M_y, \\
F_y a_x - F_x a_y + 0\ a_z &= M_z.
\end{aligned} \tag{6.6}
$$

The determinant of this system of linear equations is zero. It therefore determines a straight line – the acting line of the force vector $\boldsymbol{F}$ – with a_x, a_y and a_z as the coordinates of a point moving along it. The point of force application can, therefore, easily be found by intersecting the acting line with e.g. the surface of the top plate of the dynamometer or force plate.

In the general case, not only forces but also torques may act on the measuring system. Because a torque vector can be shifted freely in space without changing its impact on the infinitely rigid measuring system, the components T_x, T_y and T_z will be superimposed on M_x, M_y and M_z.

Therefore, equations (6.6) become

$$
\begin{aligned}
0\ a_x + F_z a_y - F_y a_z &= M_x - T_x, \\
-F_z a_x + 0\ a_y + F_x a_z &= M_y - T_y, \\
F_y a_x - F_x a_y + 0\ a_z &= M_z - T_z.
\end{aligned} \tag{6.7}
$$

The system can only measure F_x, F_y, F_z, M_x, M_y and M_z (see (6.1) and (6.2)). The components M_x, M_y and M_z of the moment vector $\boldsymbol{M}$ contain the components T_x, T_y and T_z of the torque vector $\boldsymbol{T}$ but their magnitude is not known. For this reason the constants in (6.7) are undetermined and (6.7) not only defines one straight line but all straight lines in space which are parallel to the acting line of $\boldsymbol{F}$. Therefore, the position of $\boldsymbol{F}$ can not be determined from the 6 components measured by the system. Only the direction of $\boldsymbol{F}$ is known in the general case, too.

The practical significance of the this analysis is that, in general, it is impossible to tell if a moment measured by a measuring system has been caused only by an eccentric force or only by a torque or by a combination of both.

Yet in certain types of applications it is possible to determine the point of force application. This is always the case when the measuring system is a force plate on whose working (measuring) surface only compression forces are exerted (e.g. by walking, running, driving, rolling, etc. across it). Usually the working surface of such a force plate is parallel to the xy-plane, set off by a_{z0} from the origin of the coordinate system. Figures 6.29 and 6.30 show such force plates which are used in biomechanics (measuring ground reaction forces) and in vehicle engineering (measuring tire loading).

Provided that only compression forces ($F_z > 0$) or – more precisely – no tension forces can be exerted on the working surface, no torque components T_x and T_y can act, i.e.

$$T_x = T_y = 0. \tag{6.8}$$

and equations (6.7) become

$$\begin{aligned} 0\ a_x + F_z a_y - F_y a_z &= M_x, \\ -F_z a_x + 0\ a_y + F_x a_z &= M_y, \\ F_y a_x - F_x a_y + 0\ a_z &= M_z - T_z. \end{aligned} \tag{6.9}$$

The distance a_{z0} between the working surface and the origin of the coordinate system is known. Setting $a_z = a_{z0}$, the coordinates a_x and a_y of the point of force application can be calculated from the first 2 equations directly as

$$\begin{aligned} a_x &= \frac{F_x a_{z0} - M_y}{F_z}, \\ a_y &= \frac{F_y a_{z0} - M_x}{F_z}. \end{aligned} \tag{6.10}$$

Entering a_x and a_y so obtained into the third equation allows to calculate the torque (such a torque can be applied in the form of a force couple, i.e. forces applied by friction to the working surface)

$$T_z = M_z - F_y a_x + F_x a_y. \tag{6.11}$$

In this special case of a force plate just described, we have again 6 quantities that can be determined, namely F_x, F_y, F_z, a_x, a_y and T_z.

So far we have assumed the base and top plates of the dynamometers and force plates to be infinitely rigid, an assumption which the all formulae derived are based on. In practice these plates have only a finite rigidity and will bend under load, which may cause measuring errors. This can be explained with Fig. 6.19.

If the force F_z acts at a distance x from the center of the top plate, the location of the point of force application can be derived (assuming an infinitely rigid top plate) from the measured reaction forces F_{z1} and F_{z2}. The distance x calculates as

$$x = \frac{F_{z1} - F_{z2}}{F_z} l. \tag{6.12}$$

In reality the plate is not infinitely rigid but will bend under the action of F_z and the stress distribution across the 2 sensors will become asymmetric. This means that the reaction force is not – as assumed in developing the formulae – in the center of the sensor but shifted by ε_1 and ε_2 towards the center of the plate. The effective lever arms of F_{z1} and F_{z2} relative to the center of the plate are thus changed and the distance x' obtained for defining the location of F_z is different from in the ideal case, namely

$$x' = \frac{(F_{z1} - F_{z2})l - F_{z1}\varepsilon_1 + F_{z2}\varepsilon_2}{F_z}. \tag{6.13}$$

The determination of the location of the force F_z is wrong by Δx whereby

$$\Delta x = x - x' = \frac{F_{z1}\varepsilon_1 - F_{z2}\varepsilon_2}{F_z}. \tag{6.14}$$

The error Δx is only zero for $x=0$, i.e. the point of force application is in the center, for reason of symmetry, because $\varepsilon_1=\varepsilon_2$ and $F_{z1}=F_{z2}=F_z/2$. Equally, $x=0$ for $x=l$ and $x=-l$, because F_z then loads one of the sensor centrically while the other one remains unloaded. Even that is not exactly true because the loaded sensor will yield under the centric load and that side of the plate will move down by Δz. The tilt angle α (with $\tan\alpha=\Delta z/2l$) results again in a slightly asymmetric stress distribution across the sensor. This shifts the location of the force determined slightly outward.

These types of errors can usually be minimized by appropriate design of the measuring system. Obviously these errors are smaller when the sensors are at a larger distance from each other and the contact areas (load bearing areas) with the sensors are kept as small as possible. The top plate should be made as rigid as possible. A design with ribs or a cellular structure (e.g. honey comb) helps to minimize the mass in order to achieve the highest possible natural frequency.

In practice most of these errors are eliminated, at least partially, through calibrating, i.e. by choosing the loading of the system during calibration to be as close as possible as during the later measurements.

These example should suffice to point out and illustrate the particular aspects to be considered in multicomponent measuring systems. As there are few publications on this special field [Gautschi 1972 and 1978], manufacturers of multicomponent sensors should be contacted directly for more information.

6.8 Calibration of Force Sensors

Force sensors are calibrated by applying an exactly known force in an exactly known direction and recording the corresponding output of the sensor (see definitions in 4.3.1.2).

The most accurate – but also the most laborious – calibration is obtained by using precisely known weights as reference (so-called "dead-weight calibration"). This method can only be used for sensors which have a static or at least a quasistatic response, because weights must be placed manually or mechanically on the sensor. Therefore, only sensors with quartz transduction elements, connected to a charge or electrometer amplifier, can be calibrated that way.

Dead-weight calibration is often used only for checking in situ, verifying a few calibration points. Practice has shown that it is better to lift off the weight previously placed on the sensor rather than to put it on for calibration. By first placing the weight on the sensor, resetting the charge amplifier to zero, then to operate and then lifting off the weight, one obtains a clean calibration curve showing the "negative calibration step" without any jitters, if done skillfully. For routine calibrations hydraulic presses are most widely used. The force is applied via a specially calibrated reference sensor – usually a "working standard" – on the sensor, dynamometer or force plate to be calibrated. As reference, specially selected and calibrated quartz force sensors (e.g. force links as shown in Fig. 6.8) can be used. For highest accuracy, precision strain gage force sensor are to be preferred. Not only do they have a true static response but their measuring uncertainty is also smaller (typically less than ±0,1 % FSO). The charge amplifiers used to record the output of the sensor to be calibrated and the amplifier for the reference sensor have to be calibrated electrically (see 11.8).

In order to have a clearly defined point of force application, a steel ball between 2 tungsten plates or a cylinder with a hemispherical tip is commonly used to introduce the force at the precise location of the contact point between sphere and plane.

Calibrating multicomponent force sensors is particularly demanding, especially the correct determination of cross talk. For measuring a cross talk of e.g. 0,1 % with an error of less than 10 %, the true direction of the calibration force that is applied must coincide to within 20" (0,01 %, i.e. arc tan 0,0001 = 20") with the required true direction! This is difficult to achieve in a hydraulic press because even when using a steel ball, the angle of friction in the contact point will be greater than 20". In addition, deformations in the frame of a hydraulic press may easily lead to greater uncertainties in the true direction of the applied force. Only dead-weight calibration is more accurate. This requires that the sensor (or dynamometer or force plate) can be positioned in such a way as to have the sensitive axis in a vertical position in space. With 3 axes to be calibrated, the sensor must be repositioned for each direction (Fig. 6.36). The force can then be applied either by placing the reference weights directly on it or be using a yoke, which allows the calibration weight to be hung below. Whatever method is used, multicomponent calibrations call for highly specialized equipment and experienced operators (Fig. 6.37).

Torque should preferably be calibrated by applying a force couple or through a torsion bar. Force couples can be produced e.g. with pulling ropes or wires. Less satisfactory is the method of using an eccentric force, because then not only a torque is applied – as desired – to the sensor but a force, too. A cross talk from the force into the torque can falsify the torque calibration.

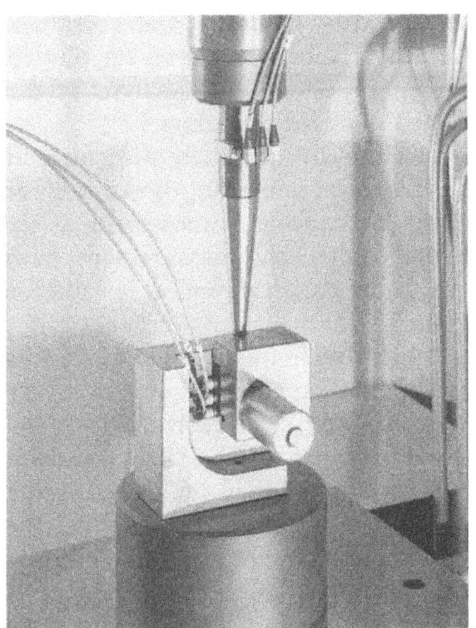

Figure 6.36 Calibrating 3-component force sensors. The sensor of Fig. 6.13 is mounted under a standard preload in a special calibration rig that can be set in all 3 positions under a hydraulic calibration press (Courtesy of Kistler)

Figure 6.37 3-component force calibration system. Calibration forces of up to 100 kN in x- and y-directions and up to 200 kN in z-direction can be applied singly or in any combination on the sensor to be calibrated (Courtesy of Kistler)

Sensor characteristics such as sensitivity, linearity, hysteresis and cross talk obtained from calibration used to be derived from calibration curves e.g. plotted with y-x-recorders. However, digital capturing, processing and analysis of the calibration data has become now the established standard procedure.

Calibrating force sensors dynamically is technically extremely difficult and requires sophisticated and costly equipment. Moreover, calibrating dynamically at the high frequencies that piezoelectric sensors are capable of measuring at are often just impossible to do. Years of practice and comparative measurements have shown, however, that the sensitivity obtained from static calibration is valid for dynamic measuring, too.

There is little sense in trying to determine the natural frequency of force sensors such as load washers. Depending on the method applied, results can vary considerably. Of more practical value is the rigidity – the "spring constant" – of the sensor. In simple measuring setups, the natural frequency can be calculated from the attached mass and the rigidity. Best results are obtained by analyzing the dynamic behavior of the measuring setup after mounting and preloading the sensor. Exciting the structure by tapping it with an instrumented impulse hammer (see 9.10) and correlating the measured input from the hammer with the output of the force sensor by performing an FFT-analysis usually gives good results. Another approach is to use modal analysis, especially with measuring systems that have a complex structure (see 9.10).

7 Strain Sensors

7.1 Quantity and Units of Measurement

The quantity "strain" – in this context meant to be "linear strain" – in an elastic body is defined as the ratio $\varepsilon = \Delta l / l_0$, where Δl is the increase in length and l_0 is the length in a reference state to be specified. The unit for strain in the SI (Système International) is "one", i.e. $1\,\varepsilon = 1 = 1\,\text{m/m}$. In practice, the "unit" for strain is called "strain" and the symbol ε is used. Usually, strain is in the order of µm/m, i.e. 10^{-6}, and therefore, the unit "µε" (microstrain) is most commonly used.

Strain must not be confused with deformation. Although strain and deformation are closely related they are 2 distinctly different phenomena. To illustrate this let us consider a bar having a length $l = 1\,\text{m}$ and a cross section $A = 1\,\text{cm}^2$, made of a material with a modulus of elasticity $E = 20\,\text{MN/cm}^2$ (steel). If a tensile force $F = 20\,\text{N}$ is applied, the length of the bar will increase by $\Delta l = (F \cdot l)/(A \cdot E)$, i.e. $\Delta l = (20\,\text{N} \cdot 1\,\text{m})/(1\,\text{cm}^2 \cdot 20\,\text{MN/cm}^2) = 10^{-6}\,\text{m} = 1\,\mu\text{m}$. This is the deformation of the bar, i.e. its length has increased by 1 µm.

The strain induced in the bar is $\varepsilon = \Delta l / l_0 = 10^{-6} = 1\,\mu\varepsilon$ and is the same all along the bar. This is not so for the deformation: The deformation e.g. at half the length of the bar (at 0,5 m) is not 1 µm but only 0,5 µm, however the strain is the same, i.e. 1 µε!

7.2 Working Principle of Piezoelectric Strain Sensors

Piezoelectric strain sensors are actually force sensors of special design. Strain is inherently associated with deformation. A force sensor is always deformed when a force is acting on it. If a force sensor is attached to a structure in such a way as to undergo the same deformation – or a constant portion of that deformation – it will yield a signal proportional to the strain. The strain sensitivity of the sensor is usually determined in a standard setup by calibration or determined in situ, then usually calibrated directly as a function of the force acting on the structure, the most common application of such sensors.

There are 2 types of strain sensors: surface strain sensors that are attached to the surface of a structure and so-called strain pin sensors that are inserted into prepared mounting bores to measure strain within the structure. Obviously such sensors can only be used on massive structures of sufficient cross-section allowing the mounting bore, possibly with a mounting thread, to be made. The main application of both sensor types is to measure – via the strain – small load changes in heavy structures such as columns or frames of presses.

An important advantage of quartz strain sensors is their exceptionally high strain sensitivity of up to 100 pC/$\mu\varepsilon$ which corresponds to over 100 V/$\mu\varepsilon$ when connected to a good charge amplifier – a value more than one million times greater than what is possible with wire strain gages. Therefore even minute strain resulting from extremely small load changes in heavy structures can be measured easily – the threshold of such sensors can go below 0,000 1 $\mu\varepsilon$ – which is far below the threshold of wire strain gages.

Although mainly used for measuring strain dynamically, most piezoelectric strain sensors have transduction elements of quartz which give excellent quasistatic response, too (11.5.6).

7.3 Surface Strain Sensors

Surface strain is traditionally measured with strain gages. Strain gages are thin and light and therefore can be attached also e.g. to thin plates or structures. On a heavy structure (at least about 2 cm thick), piezoelectric surface strain sensors can be used, too.

The cross section of a typical piezoelectric surface strain sensor (Fig. 7.1) is shown in Fig. 7.2. The sensor housing (1) is mounted with a single M6-bolt onto the surface of the structure which has previously been fine-machined at the measuring location. The contact with the finished surface of the structure is made by a contact lip (3) and a small ring-shaped contact face (4). A pair of quartz disks (5), cut for the shear effect serves as transduction element. After tightening the bolt (2) by a prescribed torque, the charge amplifier connected to the output of the

Figure 7.1 Surface strain sensor (Courtesy of Kistler)

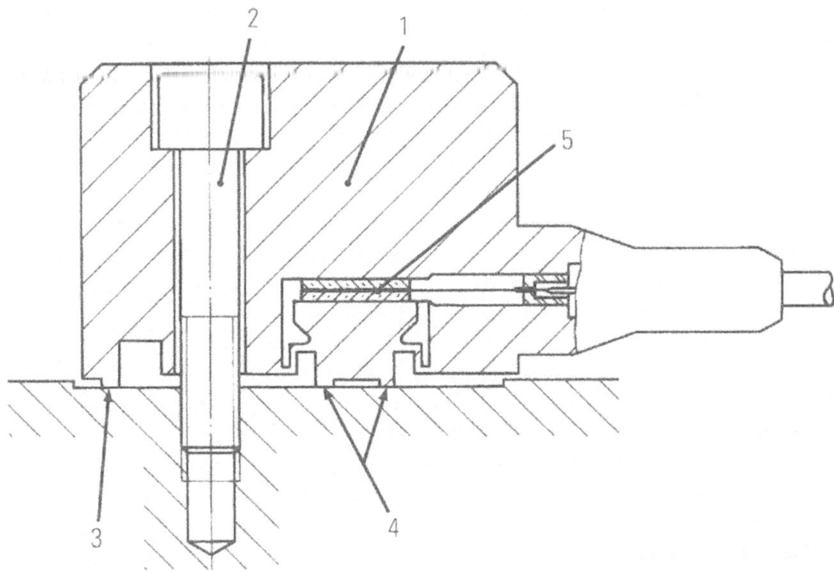

Figure 7.2 Cross-section of the surface strain sensor shown in Figure 7.1. (Courtesy of Kistler) 1 housing, *2* M6-bolt, *3* contact lip, *4* ring-shaped contact face, *5* pair of shear-sensitive quartz plates

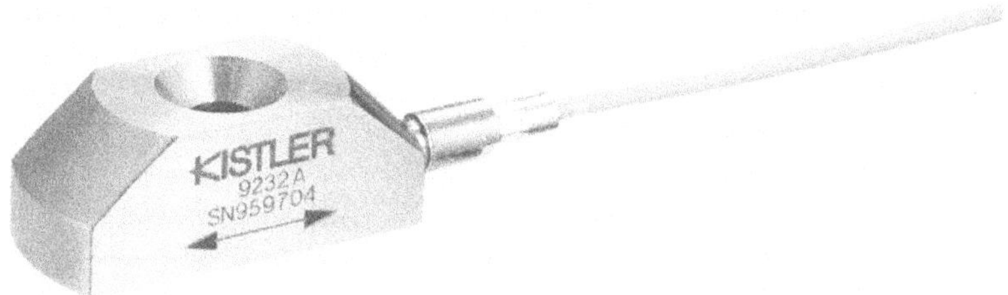

Figure 7.3 Surface strain sensor with high sensitivity of 300 pC/$\mu\varepsilon$ (Courtesy of Kistler)

sensor is reset to zero and now positive or negative strain can be measured in the direction in which the sensor is mounted. If the specified tightening torque of the bolt (2) has been correctly applied and the mounting surface has been machined to the specified surface quality, such a sensor has the advantage that it is "overload proof". If the strain in the structure becomes excessive – close to the overload of the sensor – the sensor simply slips with its contact faces on the mounting surface on the structure.

This type of strain sensor has a rather high mass (150 g for the sensor shown in Fig. 7.1) resulting in a certain acceleration sensitivity (0,6 $\mu\varepsilon/g$ for that sensor) which must be taken into account when mounting such a sensor on a vibrating structure. The sensor in Fig. 7.3 is lighter (50 g) and of low-profile design, which results in an acceleration sensitivity of only 0,03 με/g, while it offers the very high strain sensitivity of 300 pC/με.

The measuring ranges of piezoelectric surface strain sensors go up to ±300 με for bidirectional strain and 0... 600 με for unidirectional strain. The operating temperature range typically is 30...100 °C.

7.4 Strain Pins for Internal Strain

For measuring internal strain in a structure, pin-shaped sensors that mount in a specially prepared bore – similar to a "Rawlplug" – can be used. Called a measuring pin, such a sensor can be sensitive to transverse or to longitudinal strain. Longitudinal strain pins measure strain in the direction of their axis while transverse strain pins measure in one direction normal to their axis, depending on the position in which they are mounted [Cavalloni 1990 and 1991].

7.4.1 Longitudinal Strain Pins

For measuring internal strain in a structure normal to its surface, i.e. in the direction of the measuring bore made, so-called longitudinal strain pins are used. Figure 7.4 shows the design schematically, which essentially is that of a force sensor. The quartz elements cut for the transverse effect make the sensor sensitive to compression in the direction of its axis. The sensor is inserted into a measuring

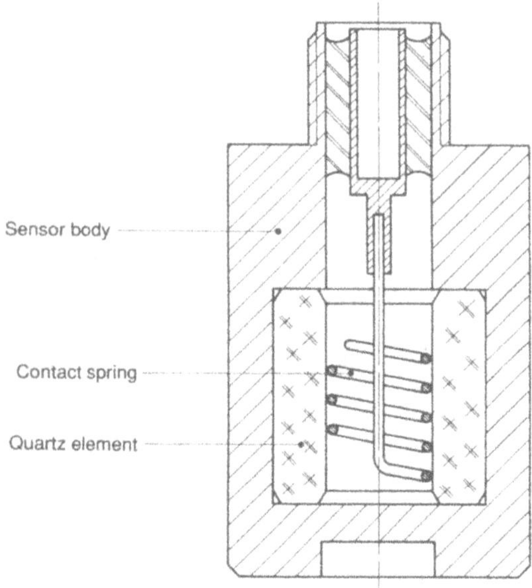

Figure 7.4 Cross-section of a longitudinal strain pin (Courtesy of Kistler)

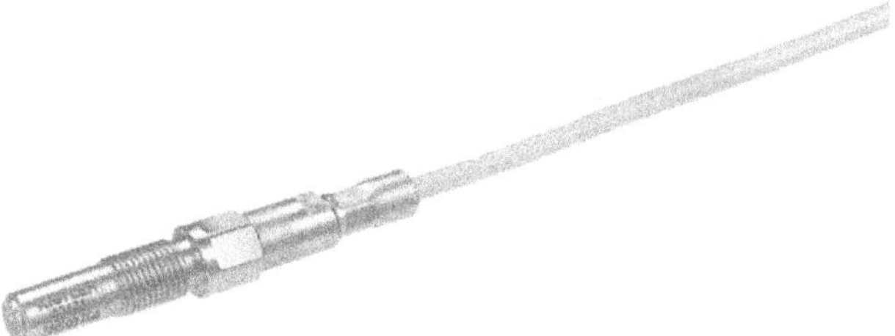

Figure 7.5 Longitudinal strain pin with an M5-thread, requiring an access bore of only 7,5 mm diameter (Courtesy of Kistler)

bore slightly larger in diameter than that of the sensor, allowing free axial movement. The bottom of the bore must be machined flat and normal to the axis of the bore. The sensor (Fig. 7.5) is inserted and compressed axially by tightening it in the threaded bore. This preload is necessary to impress the strain of the test object on the sensor, i.e. to couple it mechanically with the object to be measured. Usually the preload is chosen to compress the sensor to 50 % of its range, allowing to measure strain in both directions. If the strain to be measured is only extension or compression, the preload is chosen to compress the sensor to its full range or to only a very small fraction of its range.

Two applications serve to illustrate the measuring possibilities with longitudinal strain pins.

7.4.1.1
Measuring Tie Bar Loading

The loading of tie bars, e.g. of a plastic injection molding machine, can easily be measured and monitored with longitudinal strain pins. As shown in Fig. 7.6, the measuring bore for the sensor is made in the center axis of the tie bar. The smallest sensors required a bore of only 7,5 mm diameter. The diameter of tie bars is usually more than about 80 mm, which means that the bore reduces the cross-section - and the load bearing capacity – of the tie bar by less than 1 %. The bore can be made as deep as required for the sensor to reach an area where the stress and hence the resulting strain are well homogeneous, i.e. sufficiently away from the end of the tie bar and thus influenced by the the nut and the clamping force on the bearing plate.

The sensor, provided it has been calibrated before (see 7.6) will measure the strain to an accuracy which depends mainly on the difference between the calibration setup and the actual situation in the tie bar. The error in measuring the actual strain is usually below about 5 %. In practice, it is rarely necessary to know the actual strain precisely. Rather an in-situ calibration is made by measuring the clamping force of the machine with a force sensor temporarily inserted at the location of the injection mold and correlating it with the output obtained from the strain sensor.

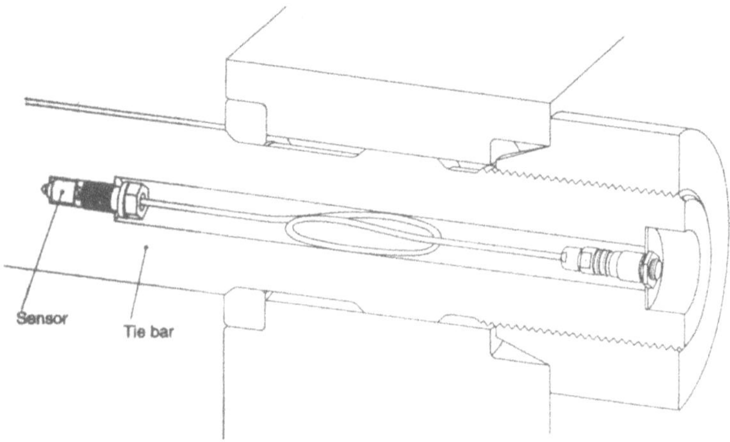

Figure 7.6 Cross-section of a tie bar, fitted with a longitudinal strain pin in a bore along its axis (Courtesy of Kistler)

There are 2 main benefits obtained from strain sensors installed in all tie bars (usually 4 per machine). First, the tie bars can be "balanced", i.e. by adjusting the nuts on the tie bar ends, equal loading – or symmetrical loading – of all tie bars can be achieved. Second, during operation of the machine the tie bar loading can be continuously monitored (note: this is not a static measurement, because the machine operates in cycles, lasting at most a few minutes). This allows to prevent damage to the machine and to the mold.

7.4.1.2
Measuring Pressure Variations in Thick-Walled Pipes

Measuring small pressure variations in thick-walled pipes or nozzles is a problem found e.g. in the nozzles of injection molding machines (Fig. 7.7). A standard pressure sensor is not suitable because the flat diaphragm of the sensor can not be matched with the cylindrical contour of the bore, leaving dead spaces in which melt may accumulate and solidify, leading to measuring errors. Also a diaphragm of the sensor may be worn by abrasive (e.g. glass-filled melts) or corrosive melts.

The extremely high sensitivity of quartz strain pins offers a solution where the sensor does not touch the melt. Figure 7.8 shows a cross-section of a thick-walled pipe with 2 longitudinal strain pins, one mounted to measure radial strain $\varepsilon_r = \Delta L_r/L$, the other tangent strain $\varepsilon_t = \Delta L_t/L$. Assuming a constant pressure outside the pipe, we obtain from the theory of elasticity for the radial strain $\varepsilon_r = \Delta L_r/L$ as a function of the pressure change Δp in the pipe and the temperature variation ΔT

$$\frac{\Delta L_r}{L} = -\frac{1}{E}\frac{r_i^2}{r_a^2 - r_i^2}\left[(1+\mu)\frac{r_a^2}{r_2(r_2+L)} - (1-2\mu)\right]\Delta p + \alpha\Delta T. \tag{7.1}$$

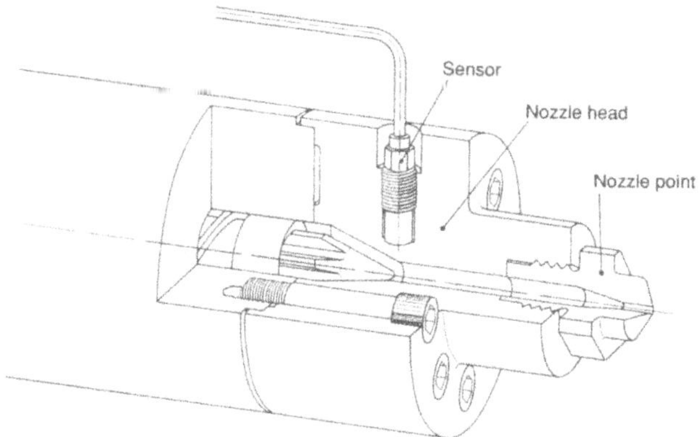

Figure 7.7 Longitudinal strain pin mounted in the nozzle of a plastic injection machine for measuring melt pressure indirectly, i.e. without touching the melt (Courtesy of Kistler)

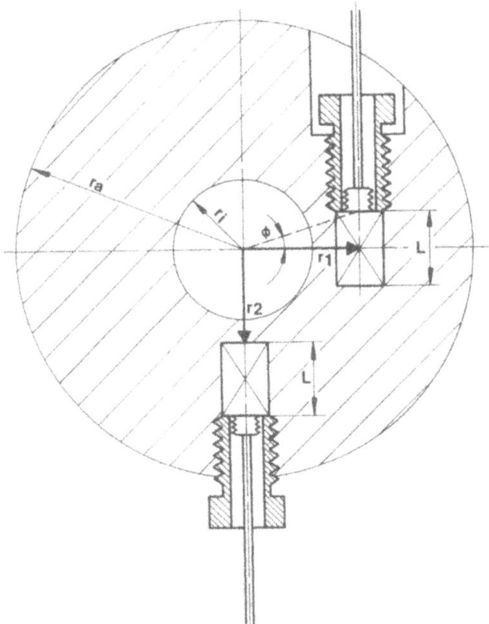

Figure 7.8 Cross-section of a thick-walled pipe with 2 longitudinal strain pin mounted for measuring the inside pressure (Courtesy of Kistler)

where E is the modulus of elasticity of the pipe material, μ the Poisson's ratio, r_i and r_a the inside and the outside radius of the pipe respectively, r_2 is the distance of the sensor front from the center of the pipe, and α is the thermal expansion coefficient of the pipe material.

For the tangent strain $\varepsilon_t = \Delta L_t / L$ we get by analogy

$$\frac{\Delta L_t}{L} = \frac{1}{E} \frac{r_i^2}{r_a^2 - r_i^2} \frac{2r_1}{L} \left[(1+\mu) \frac{r_a^2}{r_1^2} \sin\phi + (1-2\mu) \ln\, tg\left(\frac{\phi}{2} + \frac{\pi}{4}\right) \right] \Delta p + \alpha \Delta T. \tag{7.2}$$

with $\phi = \arctan(L/2r_1)$ and r_1 the distance of the sensor axis from the pipe center. For small angles ϕ, the approximation

$$\ln\, tg\left(\frac{\phi}{2} + \frac{\pi}{4}\right) \cong \sin\phi, \tag{7.3}$$

simplifies equation (7.2).

Increasing internal pressure in the pipe produces a compressive strain in the radial sensor, but a tensile strain in the tangent sensor. Temperature changes, however, result in the same type of stress in both sensors, i.e. both compressive or both tensile.

For measuring the pressure change with least influence by the temperature change, the difference of the output signals of the 2 sensors must by taken. This means that each sensor would have to be connected to a charge amplifier and then the difference of the respective output voltages be taken. A simple trick simplifies the setup considerably. Since the outputs of two identical piezoelectric sensors can simply be switched in parallel at the input of a charge amplifier to obtain the sum of both signals (see 11.6), one of the two sensors is fitted with the piezoelectric element for the opposite polarity (sensors with positive or negative sensitivity are commercially available from the manufacturers). Thus the "sum" of the two outputs becomes the desired difference and only one charge amplifier is sufficient. The strain resulting form the temperature change being the same in radial and tangent direction, the pressure change can be measured unperturbed by temperature change.

Obviously, the two strain sensors installed and switched in parallel to a single charge amplifier must be calibrated for their combined output in response to the pressure change to be measured by comparing e.g. their output with the output of a standard pressure sensor used in the calibration setup. For obtaining the best linearity, it is advisable to make $r_1 \approx r_2$.

If temperature compensation is not needed – such as in dynamic processes with short cycling times – one sensor may be sufficient. When a single sensor is used it should be measuring the tangent strain because it gives a higher sensitivity to pressure as can be seen from (7.1 and 7.2).

7.4.2 Transverse Strain Pins

Transverse strain pins measure strain normal to their axis. They allow to measure strain deep inside a solid structure such as a frame of a column of a press. Fig. 7.9 shows a transverse measuring pin. Inserted into a mounting bore

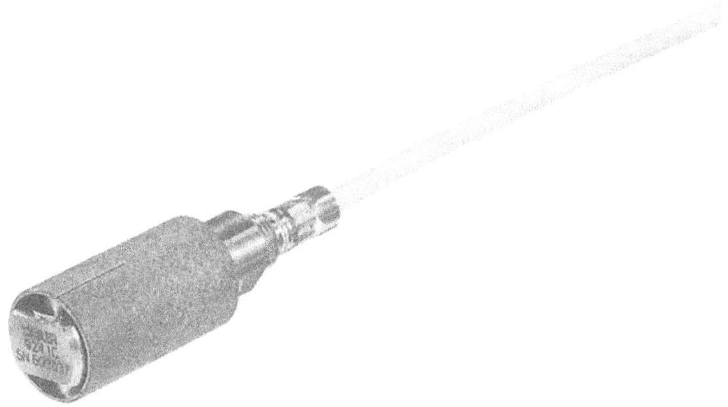

Figure 7.9 Transverse strain pin requiring a mounting bore of 10 mm diameter (Courtesy of Kistler)

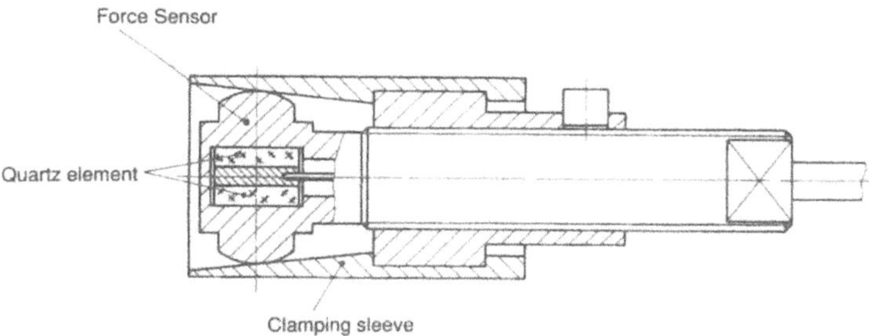

Figure 7.10 Cross-section of a transverse strain pin (Courtesy of Kistler)

prepared to specified tolerances the sensor is first positioned with its sensitive axis in the desired direction and then clamped tight by expanding its diameter by means of a tapered clamping sleeve into which the sensor element is pulled by a screw (Fig. 7.10). As transduction element 2 quartz disks cut for the longitudinal effect are used in an arrangement similar to the load washers described in 6.2.1.

This type of sensor can easily be positioned and repositioned in a bore. That makes it possible to e.g. obtain a detailed strain profile in a heavy structure. Also, the optimal position in a given application can be found. i.e. the location and direction having the highest strain sensitivity or the best linear relationship between the force to be monitored in the structure and the measured strain.

7.5 Omnidirectional Strain Sensors

A highly sensitive strain sensor that reacts to strain in any direction in its contact area with the measured object is simply pressed onto the surface. The cross section is shown in Fig. 7.11.

The piezoelectric surface-strain sensor has a circular sensing/transduction element, consisting of a thin steel diaphragm with a layer of piezoelectric ceramic material permanently bonded to one of its faces. The uncoated side of the steel disc is pressed onto the surface of the structure whose strain is to be measured. To maximize the coupling the transduction element is pressed with a soft material of a Young's modulus much lower than that of steel, resulting in a hydrostatic-like loading and thus assuring uniform pressure distribution and preventing local slippage.

The soft cushioning material inside the sensor housing allows the sensor to be simply bolted on without the need for a torque wrench, because the cushioning material is chosen of such a volume that, when the housing is pressed tightly

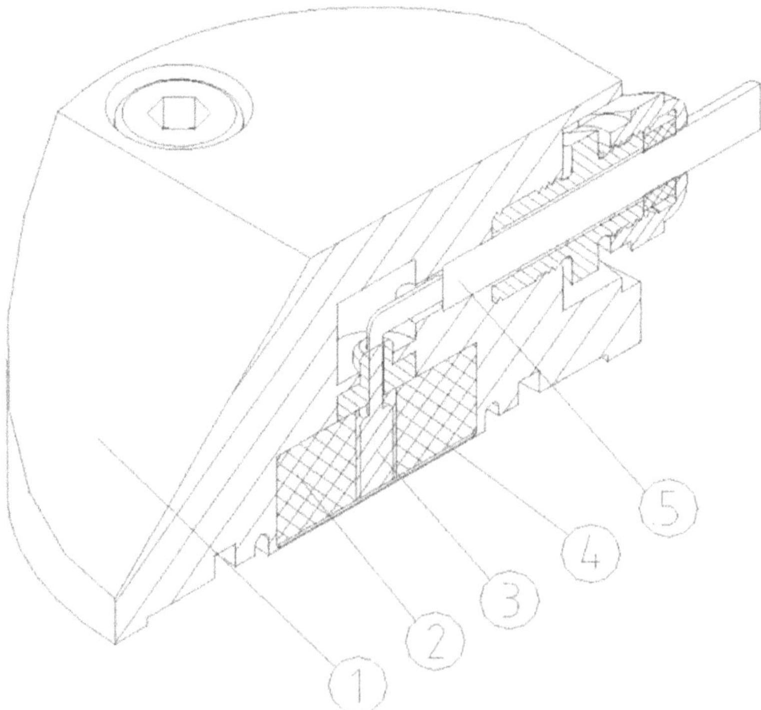

Figure 7.11 Omnidirectional strain sensor (Courtesy of Baumer sensopress) *1* housing, *2* cushioning material ("hydrostatic element"), *3* contact pin, *4* sensing/transduction element (steel diaphragm with bonded-on piezoelectric ceramic layer), *5* cable

against the machined mounting surface, the intended contact pressure is automatically obtained.

The circular transduction element gives the sensor a high omnidirectional sensitivity which can reach to around 1000 pC/µε. The range is ±500 µε in any direction wile the threshold is far below 1 nε.

7.6 Piezoelectric "Strain Gage"

A strain sensor shaped like a thick strain gage (5,1 · 15,2 · 1,8 mm) and of only 0,5 g mass is shown in Fig. 7.12. It can be applied to a surface by bonding with an adhesive and will give a sensitivity of about 50 mV/µε. The frequency range extends over 0,5 ... 100 000 Hz and the threshold is below 1 nε. It has a quartz transduction element and a built-in amplifier acting as an electrometer amplifier (see 11.3). Excited by a constant current, it provides an output signal that is insensitive to noise and external electric fields.

Another design is the so-called strain-link (Fig. 7.13). The body of the sensor contains 6 elements as described in 7.5, arranged in a straight line – the direction in which the sensor will measure – between 2 bolts that are 50 mm apart. The shape of the sensor body is such that all elements are pressed sufficiently against the structure to insure that the strain is fully transmitted through friction after tightening the bolts correctly. Mechanically the 6 elements

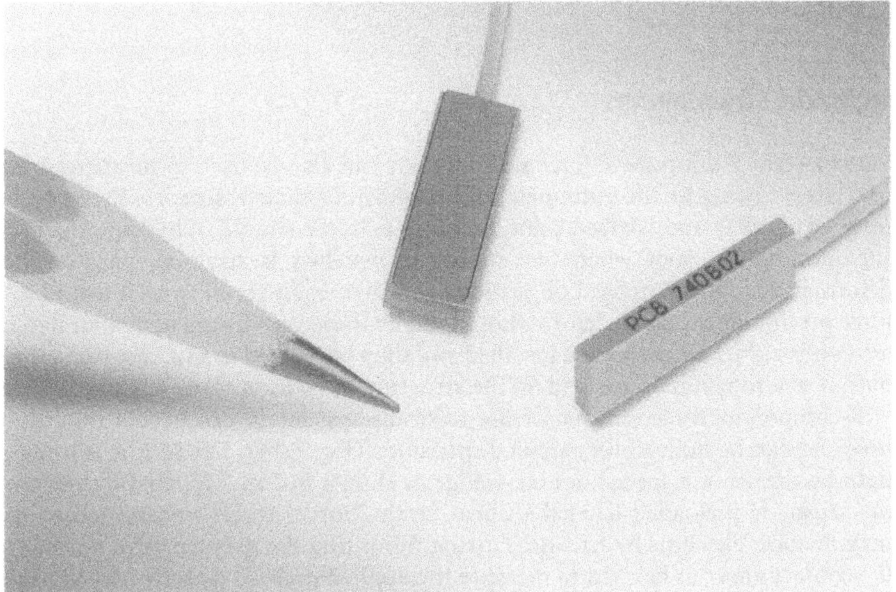

Figure 7.12 Piezoelectric strain-gage (Courtesy of PCB)

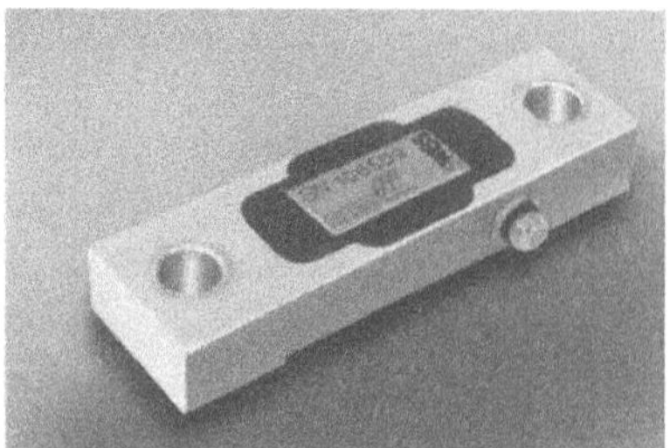

Figure 7.13 High-sensitive "Strain-link" sensor with a sensitivity of 5000 pC/με (Courtesy of Baumer sensopress)

are in parallel while electrically they are in series. This gives the sensor a very high sensitivity of around 5000 pC/με, i.e. over 5 V/με with a good charge amplifier.

The transduction elements are made of piezoelectric ceramics which means that a strong pyroelectric effect is present. Depending on the application, the sensor must be protected against temperature transients by an appropriate cover. The lower frequency limit is around 1 Hz. Quasistatic measuring is not possible and aperiodic events should not last more than about 5 s for large strains (correspondingly shorter for small strains). When measuring aperiodic events, the amplifier (which can also be built into the sensor) must be reset between events. This is usually done with a semiconductor reset switch (e.g. a J-FET, see 11.5.3)

7.7 Multiaxial Strain Sensors

Piezoelectric 3-component force sensors (6.3) can also be used to measure strain in 3 axes. Owing to the extremely high rigidity of quartz sensors, they can be inserted in slots and wedged tight, forming a "force shunt". When the material surrounding the spot where the sensor is installed is strained, part of that deformation will be imposed on the sensor which reacts similarly as it would to a force producing the same deformation. Usually these sensors are not calibrated in terms of strain. Rather they are installed and then calibrated in situ, recording their output as a function of the load on the structure to be monitored later.

3-component force sensors for use as strain sensors do not need a mounting hole and can be built to very small dimensions (Fig. 7.14). The sensor is wedged tight by means of a special double-wedge as shown in Fig. 7.15. By this wedging the sensor is preloaded to enable shear strain (force) to be transmitted to the transduction elements by friction. During mounting the axis sensitive normal to the contact area can be used to measure the applied preload directly. The amount of preloading depends on the desired measuring ranges and is usually specified by the manufacturer for various practical applications.

Figure 7.14 Miniature 3-component force sensor that can also serve as strain sensor (Courtesy of Kistler)

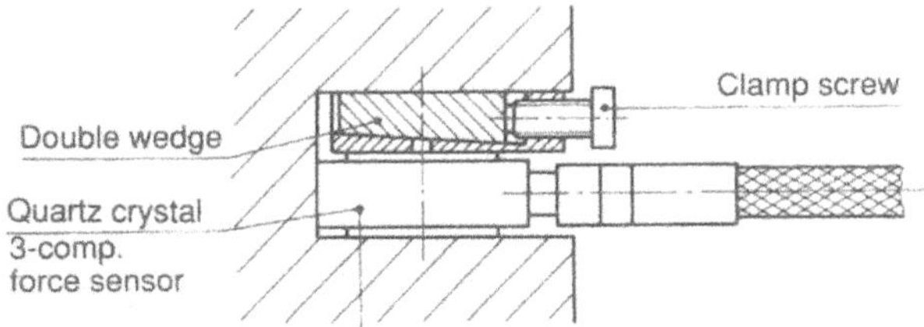

Figure 7.15 Sensor of Fig. 7.14 wedged into a slot in the measured structure to act as a 3-component strain sensor (after Kistler)

7.8 Calibration of Piezoelectric Strain Sensors

Piezoelectric strain sensors must be calibrated to determine their strain sensitivity. Usually, the strain sensor is attached to a reference solid (for surface strain sensors) or mounted in a bore in a reference solid. There are no standards as to the shape, size and material of such reference bodies. Usually each manufacturer has his own design and will refer to it in the calibration certificates.

Yet such calibrations are not absolute because the strain sensitivity so determined is valid only for the type of material (of equal Young's modulus) the test solid is made of. The strain produced in the spot on the reference solid where the sensor is calibrated is usually calculated only, sometimes verified with the help of strain gages applied around the mounting spot.

An example of such a reference solid for calibrating longitudinal strain pins is shown in Fig. 7.16. It is a cylinder whose diameter is much larger than the diameter of the sensor, mounted coaxially at about half the height of the cylindrical body. The strain is assumed to be homogenous (the force must be applied evenly distributed across the top face of the cylinder which can e.g. be achieved by interposing a thin neoprene mat between the force-introducing part and the top of the calibration solid) and is calculated from the acting force and the cross-section of the cylinder at the height where the sensor is mounted.

The sensitivity derived from such calibrations is approximate only and may serve as a guide for selecting a sensor and for determining the parameters for installing it. Piezoelectric strain sensors are mainly used to measure force acting on the structure into which they are mounted. Therefore, they are calibrated in situ directly as a function of an applied force of known magnitude. Often, the absolute

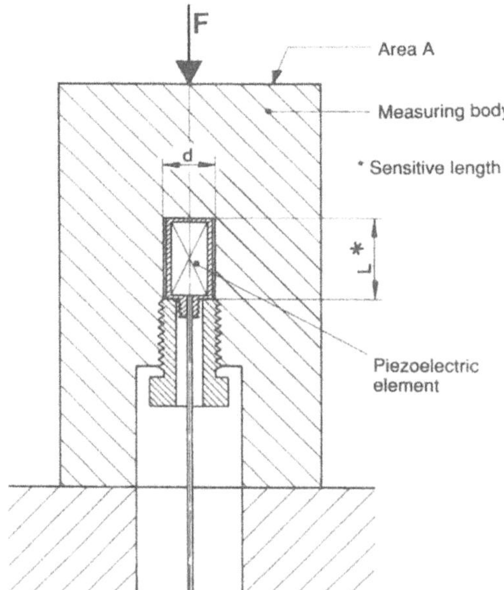

Figure 7.16 Calibration solid (steel cylinder) with a longitudinal strain pin mounted in its center bore (after Kistler)

value is not of interest, especially in process control. The sensor serves only to insure that a value found to be suitable for the process is reached each time again in subsequent cycles and to signal deviations from such values found empirically. The inherently very high reproducibility of quartz sensors responds particularly well to this requirement.

8 Pressure Sensors

8.1 Quantity and Units of Measurement

The quantity "pressure" in a fluid (a liquid or a gas) is defined as "force divided by area". The unit for pressure in the SI (Système International) is the "Pa" (Pascal), defined as $1\,\text{Pa} = 1\,\text{N/m}^2$. Unfortunately, 1 Pa is a very small pressure compared with the barometric pressure, which is about 100 kPa. Pressure encountered in most practical applications, especially in hydraulics and pneumatics, is always related to barometric (atmospheric) pressure. Because the unit "Pa" is not very convenient to use in these fields, the coherent unit "bar", defined as $1\,\text{bar} = 100\,\text{kPa}$ (exactly) is widely used and also accepted by ISO 31-3:1992(E), item 3-15.1, for use in these fields. Therefore, in this book, the "bar" is used as the working unit for pressure.

A pressure of 1 bar corresponds nearly to the barometric pressure at sea level, with the standard atmosphere defined as $1\,\text{atm} = 1{,}013\ 25\,\text{bar}$ (exactly). The unit "bar" can be combined with decimal prefixes, such as "µbar" for use in acoustics, "mbar" in meteorology and medicine, "bar" and "kbar" in pneumatics and hydraulics.

Piezoelectric pressure sensors are a special form of force sensors. They nearly always have a diaphragm which ideally has a constant effective area. The force exerted by a fluid on the diaphragm is therefore proportional to the pressure, i.e. the pressure measurement is based on a force measurement.

There are basically two types of pressure sensors: *absolute pressure* and *differential pressure* sensors (Fig. 8.1). Sensors for absolute pressure p_{abs} have a sealed chamber, containing the reference pressure p_{ref}. Because p_{abs} refers always to zero pressure, vacuum is used commonly used as p_{ref}. Therefore, the reference side of the diaphragm is exposed only to vacuum. If p_{ref} is chosen not to be vacuum, the temperature dependence of the pressure of the sealed-in gas representing p_{ref} must be taken into account. Differential pressure sensors do not measure relative to vacuum but to a reference pressure p_{ref}. If the ambient barometric pressure is used as p_{ref}, the sensor becomes a relative pressure sensor (often called *relative pressure* or *gage pressure* sensor). Here, the reference side of the diaphragm is exposed to

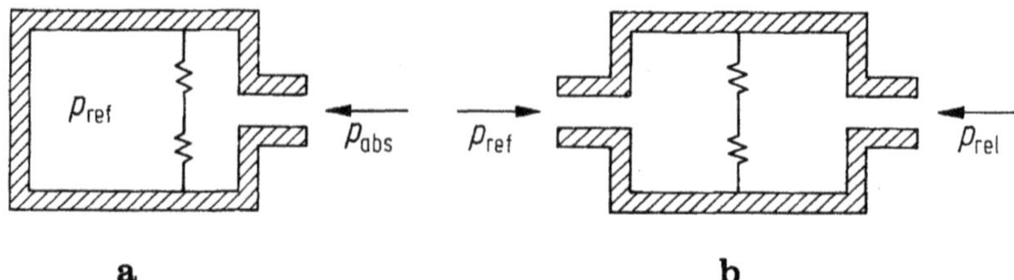

Figure 8.1 Absolute and relative (gage) pressure sensors. **a** Absolute pressure sensor, **b** relative pressure sensor

atmosphere, including humidity and possibly dirt and corroding substances. True differential pressure sensors measuring the difference between p_1 and p_2 must usually accept the same fluid on both sides of their diaphragm. Therefore the diaphragm and the access ports of the sensor must be designed to resist the fluids they are exposed to, especially when they are aggressive or corrosive.

Absolute pressure sensors can be used to measure barometric pressure or pressure in processes which require absolute pressure (e.g. chemical or biological processes). Relative (gage) pressure sensors are used to measure filling levels in reservoirs while differential pressure sensors can serve to measure the pressure drop in nozzles and orifices for determining flow.

All the applications described above require that the pressure sensor can measure statically. This is inherently not possible with piezoelectric sensors. Either the charge amplifier connected to the sensor is reset just before the measurement, which defines the pressure existing at that instant and the reference pressure for the quasistatic measurement (near DC-mode) following, or the sensor is only used to measure the dynamic relative variations (AC-mode) in the pressure (see chapter 11).

For these technical reasons, no distinction between absolute and relative pressure sensors can be made in piezoelectric sensors. Therefore we speak only of "pressure sensors" and mention just differential pressure sensors as a special design with two pressure ports allowing to measure dynamic and quasistatic pressure differences.

Depending on their range, pressure sensors are often classified as low-pressure sensors (up to a few bar), as pressure sensors for general application (up to several hundred bar) and as high-pressure sensor (over 1 kbar). Low-pressure sensors which measure in the region of µbar are also called piezoelectric microphones.

Pressure sensors that work at temperatures above 250 °C are called high-temperature pressure sensors.

8.2 Design of Piezoelectric Pressure Sensors

The typical design of a standard pressure sensor is shown in Figs. 8.2 and 8.3. The piezoelectric transduction element consists of 3 quartz columns cut for the

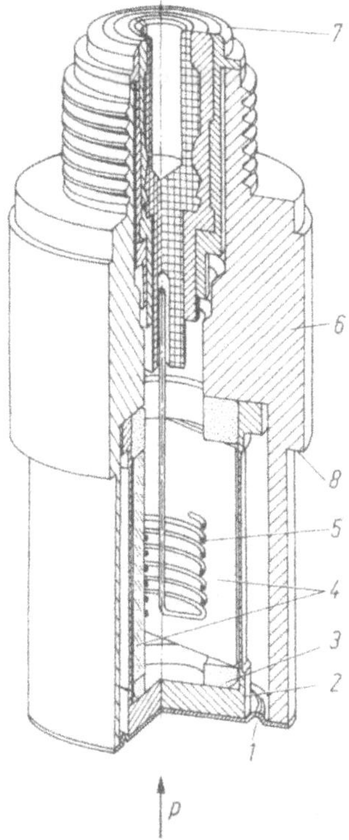

Figure 8.2 Design of a standard pressure sensor (Courtesy of Kistler)

Figure 8.3 Low-pressure sensor, using a high-sensitivity force sensor shown in Fig. 6.10 as sensor element (Courtesy of Kistler)

transverse effect (*4*), preloaded by a preloading sleeve (*5*). The front end of the sleeve acts as a force-transmitting part and the interposed plate (*3*) serves to equalize the stress distribution on the end faces of the quartz elements and to compensate temperature effects. A similar plate is found for reasons of symmetry at the other end of the quartz elements. The diaphragm (*1*) is welded with a light preload against the preloading sleeve and flush with the sensor housing (*6*), forming a hermetic seal. The sensor housing usually has a sealing shoulder (*8*), some types have a mounting thread, too (e.g. as shown in Fig. 8.6). On the other end is the electric connector (*7*), usually of the coaxial type.

Piezoelectric elements exploiting the transverse effect yield the electric charge on their mechanically unloaded faces, which are therefore coated under vacuum with a metal to form an electrode collecting the complete electric charge. In the design shown, only the flat face needs to be coated. A helical-shaped spring (*6*) contacts this electrode and carries the electric charge to the connector (*7*). The other, cylindrical, side face does not need to be coated with an electrode because the charge is picked up by the preloading sleeve through capacitive coupling.

The pressure p exerted by the measured fluid on the diaphragm (*1*) is transmitted according to the active area of the diaphragm as a proportional force on to the transduction element.

The diaphragm is the critical part of a pressure sensor and usually determines the life time of the sensor. Although the diaphragm is usually very thin (less than 0,1 mm) the minute deformations occurring during pressure cycles still heavily strain it mechanically. The diaphragm has to fulfill several contradictory functions: it should prevent the fluid from penetrating into the sensor housing (e.g. not to exert radial pressure on the preloading sleeve, which not only would lead to measuring errors but could damage the sleeve and the transduction element), it should offer reliable resistance against aggressive fluids, it should be very flexible and ideally elastic in order not to impair the linearity of the sensor, and finally, it should support temperature changes with a minimum of influence on its elastic behavior while having an equalizing effect on the temperature distribution. These conflicting requirements have lead to a large number of diaphragm designs.

High corrosion resistance and at the same time a wide operating temperature range in many applications limit the choice of materials for the diaphragm. Furthermore pressures above 1 kbar quickly reach or exceed the elastic limit of many otherwise suitable materials. The elastic limit is usually reached first in the area where the diaphragm bridges the gap between sensor housing and front of the transduction element. Similarly in the sealing area of the sensor the contact pressure must be higher than the measured pressure in order to obtain a reliable seal through preloading the sensor in the mounting bore.

8.3 Low-Pressure Sensors

Low-pressure sensors must have a high sensitivity. This can be achieved by using quartz elements cut for the transverse effect or piezoelectric ceramic elements.

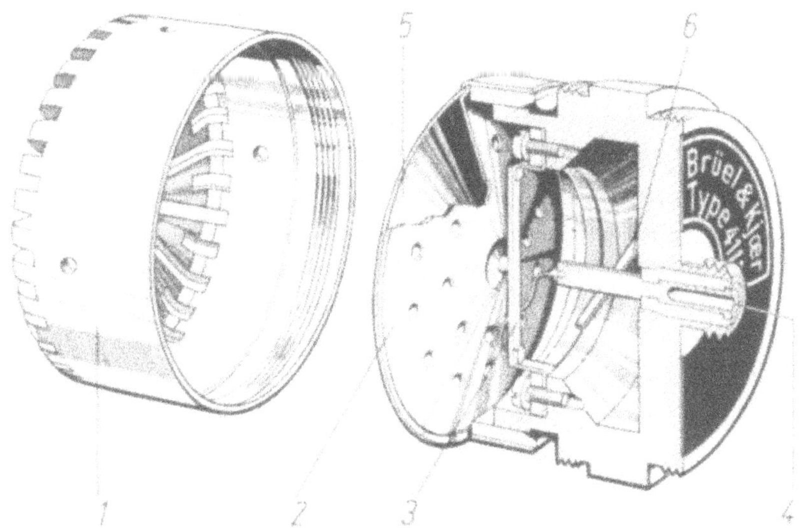

Figure 8.4 Piezoelectric microphone (Courtesy of Brüel & Kjær). *1* protective grid, *2* damping plate, *3* transduction element in the form of a bending beam made of piezoelectric ceramic, *4* connector, *5* diaphragm, 6 capillary tube for equalizing slow variations in the ambient pressure

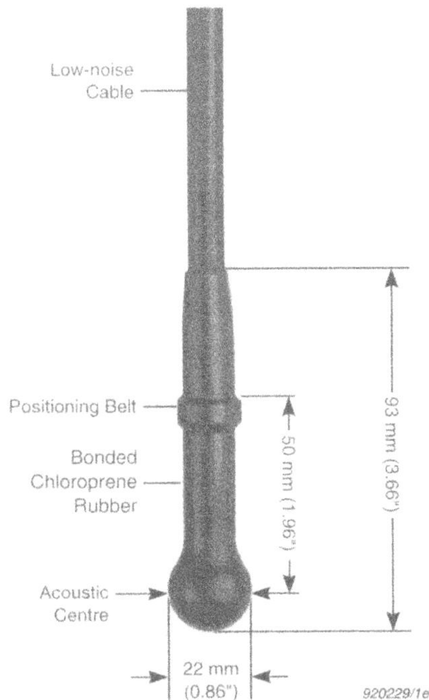

Figure 8.5 Piezoelectric hydrophone (Courtesy of Brüel & Kjær)

Also a diaphragm of large diameter increases the sensor sensitivity but leads inherently to a lower natural frequency, i.e. a more limited frequency response.

A low-pressure sensor with a range of 0...10 bar_{abs} and a threshold of about 10 μbar is shown in Fig. 8.3. The diaphragm with a diameter of about 30 mm limits the natural frequency of this sensor to about 13 kHz which is sufficient for most technical applications. Piezoelectric microphones must have a much lower threshold and still cover a frequency range as wide as possible. The microphone illustrated in Fig. 8.4 has a transduction element in the form of a bending beam obtained by sandwiching together two strips of lead-zircon-titanate working by the transverse effect. Under bending, one element is stretched, the other is compressed. With both elements switched in parallel a high sensitivity is obtained. The threshold is much below 1 μbar and frequency response is within ±3 dB from 2 Hz to 10 kHz.

Hydrophones are pressure sensors used for measuring sound under water. They must support the high static pressure when immersed in water, yet have a very low threshold. The sensor shown in Fig. 8.5 works at depths down to 1000 m, which corresponds to 100 bar, yet has a sensitivity of 42 000 pC/bar, allowing to capture pressure variations in the μbar range. The frequency ranges covers 0,1 Hz to 180 kHz.

8.4 Pressure Sensors for General Applications

Sensors with ranges of several hundred bar are available in a large number of designs. Most of them have a flush-welded diaphragm which allows them to be mounted flush with the wall of the pressure chamber. This eliminates any dead volume and the generally high (for some types to over 500 kHz) natural frequency can be fully exploited.

There are essentially two types of basic design: sensors with a mounting thread for directly screwing the sensor into the object, which must be fitted with a mounting bore having the corresponding thread cut (e.g. Fig. 8.8), and sensors without mounting thread, which mount into bores providing only a sealing shoulder and are held in place by mounting nuts, nipples or adapters (e.g. Fig. 8.6). Both systems have their advantages and drawbacks.

For sensors with a mounting thread a precise mounting bore must be made, respecting tight tolerances for orthogonality between sealing face and axis of the thread, flatness and surface finish of the sealing surface. Otherwise the sensor may be strained during mounting and characteristics such as linearity and sensitivity could be influenced.

Sensors without mounting thread are mounted in appropriate mounting bores by holding them in place with mounting nuts or nipples Or they are first mounted in an adapter which then is mounted in a corresponding bore. Adapters require more space but have the advantage that they provide a precise, well defined mounting geometry for the sensor which insures that all their specifications will be maintained. Adapters being more massive and rugged than a sensor are easier

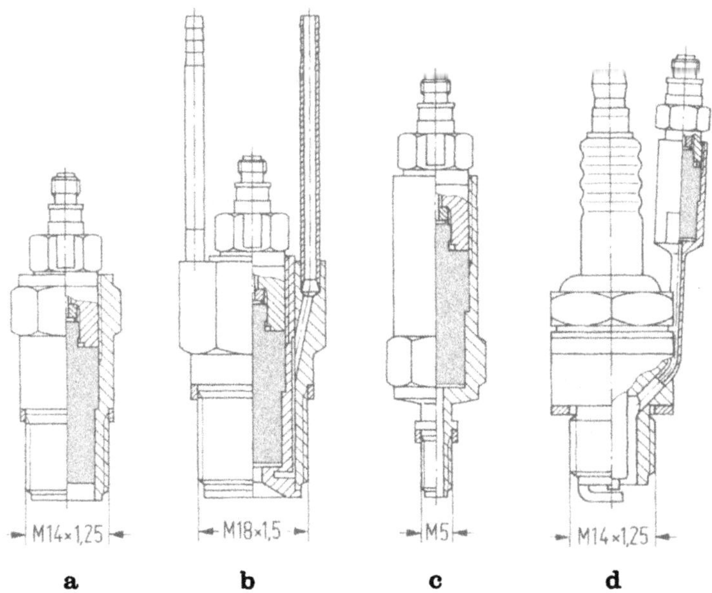

Figure 8.6 Adapters for mounting pressure sensors (Courtesy of Kistler) **a** standard adapter for e.g. the pressure sensors shown in Fig. 8.11, **b** water-cooled adapter, **c** so-called "needle adapter", **d** spark plug adapter

to handle in rough environments and require less stringent tolerances in the mounting bore. Sensors that are particularly sensitive to mounting tolerances can be first mounted in an adapter and then calibrated. As long as they are not dismounted they will keep their calibration data. A further advantage of sensors without own mounting thread is that with a small number of types and a large number of adapters (Fig. 8.6) a variety of applications can be covered.

Examples of sensors with quartz elements for general applications in standard and miniaturized version are shown in Fig. 8.7. The ranges cover 0...100 bar to 0...1 kbar, the natural frequencies are between 50 and 200 kHz. The threshold is of the order of mbar. Such sensors can be used up to about 250°C. For higher temperatures, either water-cooling or sensors with transduction elements made of piezoelectric materials with a higher temperature range (see chapter 3) must be used.

There is no general rule on what operating temperature can be attained with water-cooling. The cooling effect depends above all on the rate of flow, but also on the temperature of the inflowing water, the temperature at the measuring point and the heat transfer through the diaphragm to the sensor housing and transduction element. Water-cooling has drawbacks too, because the flow may induce vibrations which show in the sensor output. Should the water-cooling fail the sensor may be destroyed, and keeping the sensor housing at a low temperature favors condensation which can lower the insulation resistance of the transduction element and the insulation materials used inside the sensor and the connector.

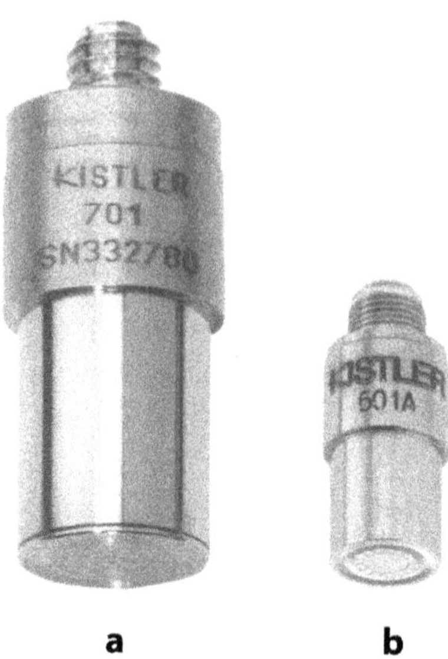

a b

Figure 8.7 Pressure sensors without mounting thread for general applications (Courtesy of Kistler) **a** standard version with diaphragm of 9,5 mm diameter, **b** miniature version with diaphragm of 5,5 mm diameter

8.5 High-Pressure Sensors

Sensors with ranges of over 1 kbar are called high-pressure sensors Their main application is in high-pressure hydraulics, in particular fuel injection pumps of diesel engines, and ballistics.

The design is essentially the same as that of standard pressure sensors. The inherent problem is that with the high pressure the sensor is intended to measure, the mechanical stress in the sensor housing and the transduction element becomes correspondingly high. For this reason the highest pressure currently possible is about 10 kbar (Fig. 8.8), because the elastic limit of the materials suitable for sensor housings (high-strength alloy steel) is of about the same magnitude. Approaching or even slightly exceeding the elastic limit during mounting and operation of the sensor means that e.g. the linearity of these sensors is not as good as with sensors for lower pressure ranges because the dependence of Young's modulus of elasticity on the mechanical stress can not be neglected any more.

High-pressure sensors must be mounted with a sufficiently high torque to insure proper sealing which makes the sealing faces the most critical parts. The high local stress in the sealing faces and correspondingly in the mounting thread causes deformations of the sensor housing which in turn influences the preload of the transduction element and can change the sensor sensitivity, i.e. the sensitivity

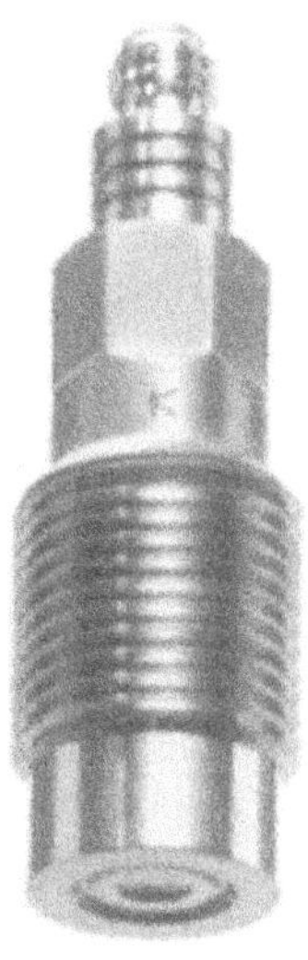

Figure 8.8 High-pressure sensor with a range of 0...10 kbar (Courtesy of Kistler)

Figure 8.9 High-pressure sensor (range: 0...2 kbar) for measuring fuel injection pressure in Diesel engines with a life of $>10^7$ cycles (Courtesy of Kistler)

depends on the mounting torque (sometimes specified as torque-dependent sensitivity). Tightening the sensor to a closely specified mounting torque is not enough because the resulting mechanical stress in the sealing face depends also on the type of lubricant (grease) used and on the surface quality of the thread.

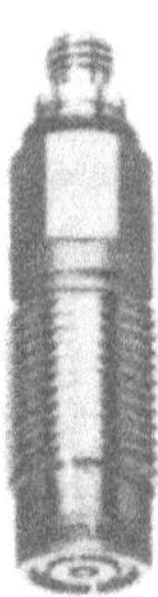

Figure 8.10 High-pressure sensor (range: 0...6kbar) for ballistics (Courtesy of Kistler)

Manufacturers of high-pressure sensors often recommend to use their special tool in preparing the mounting bore and also the type of grease to be used, besides the required torque, to maintain the original calibration values as closely as possible.

The high-pressure sensor shown in Fig. 8.9 is designed for continuous monitoring of the pressure in the fuel injection lines of diesel engines where excellent cycling life is required. Of similar construction the sensor in Fig. 8.10 used to measure the explosion pressure behind the projectile in ballistics.

The diaphragm, especially in high-pressure sensors, limits the life of a sensor, because the part that bridges the gap between the front of the transduction element and sensor housing is subjected to plastic deformation during each pressure cycle. The resulting fatigue leads to rupture after a certain number of pressure cycles. Sensors for ballistics may last a few thousands cycles only while in hydraulic applications, the sensor life may extend over several million cycles.

Various aspects of high pressure measurement techniques are described in [Peggs 1983] together with an extensive list of references.

8.6 Pressure Sensors with Acceleration Compensation

Pressure sensors are sensitive to acceleration, especially in the direction of their axis. This is because the mass of the diaphragm and transmitting plates in front of the transduction element acts as a seismic mass in the same way as in an acceleration sensor. When the sensor is accelerated or vibrated this gives a signal which is superimposed on the pressure output. Called acceleration error, this undesired signal is typically in the order of a few mbar/g. In water-cooled sensors, the acceleration error is increased by the mass of the cooling water that is contained in the part of the cooling ducts that is in front of the transduction element.

While in most applications the acceleration error can be ignored, it can severely disturb measurement where the sensor is subjected to strong vibrations while trying to measure small pressure. This led to the design of special acceleration-compensated pressure sensors.

An acceleration-compensated pressure sensor with X-cut quartz elements for the longitudinal effect is shown in Fig. 8.11, the design in Fig. 8.12. Two

Figure 8.11 Acceleration-compensated pressure sensor (design shown in Fig. 8.11) with a natural frequency of ≈500 kHz. Diaphragm: 5,5 mm diameter, mass: 1,7 g (Courtesy of Kistler)

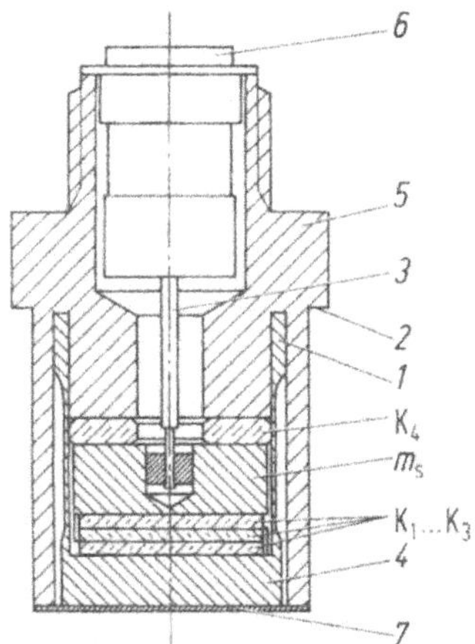

Figure 8.12 Acceleration-compensated pressure sensor (Courtesy of Kistler). *1* preloading sleeve, *2* sealing shoulder, *3* connection of the electrodes E_2, E_4 and E_6 with the connector (see Fig. 8.13), *4* pressure transmitting piece (part of preloading sleeve), *5* housing, *6* connector, *7* diaphragm, $K_1 \dots K_4$ X-quartz plates, m_s seismic mass

piezoelectric elements of different piezoelectric sensitivities are used for acceleration compensation. In Fig. 8.12, the first element consists of 3 quartz plates K_1, K_2, K_3, the second of a single quartz plate K_4.

In general the masses of the sensor parts whose inertia plays a role and the sensitivities of the two piezoelectric elements are chosen in such a way that inertial

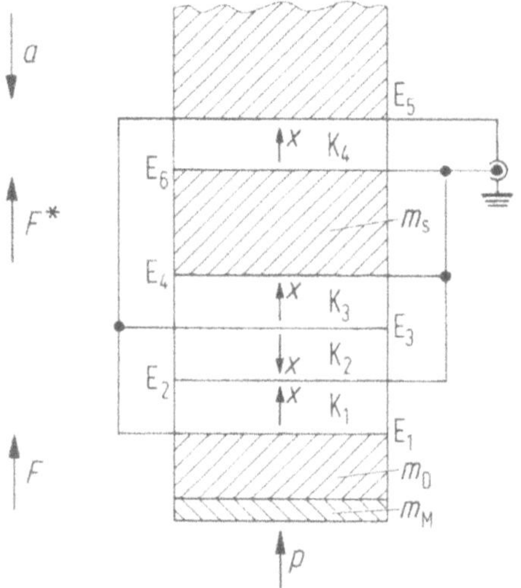

Figure 8.13 Schematic construction of the sensor in Fig. 8.12. $E_1 \dots E_6$ electrodes, F inertial force acting on $K_1 \dots K_3$ due to the acceleration a, F^* inertial force acting on K_4 due to the acceleration a, $K_1 \dots K_4$ X-quartz plates (the arrows indicate the direction of the positive crystallographic x-axes, a acceleration acting on the sensor along its axis, m_D mass of diaphragm, m_s mass for the acceleration compensation (seismic mass), p pressure acting on the sensor (measurand)

forces produce the same piezoelectric signal in both elements. Since the elements are switched in series with opposite polarities, the signals originating from acceleration cancel each other. The measured pressure, however, produces different signals in the two elements and the difference of these signals is the output proportional to the pressure acting.

In order to explain the conditions for obtaining a correct acceleration compensation, we first assume that just the masses m_M of the diaphragm, m_D of the transmission piece and m_s the compensation mass proper contribute to the compensation. Also we consider only acceleration along the axial direction of the sensor and ignore the effects of the elastic force shunts such as the preloading sleeve. The piezoelectric sensitivity of the plates K_1, K_2, K_3 is determined by the piezoelectric coefficient d_{11} and that of the quartz plate K_4 by the piezoelectric coefficient d^*_{11}.

The orientation of the plates is indicated in Fig. 8.13. The electrodes E_1, E_3 and E_5 are connected to the sensor ground while the electrodes E_2, E_4 and E_6 are connected together and serve as output electrodes, connected to the connector of the sensor. When an acceleration a is acting on the sensor, the inertial force

$$F = (m_M + m_D)a \tag{8.1}$$

acts on the piezoelectric plates K_1, K_2, K_3 and

$$F^* = (m_M + m_D + m_S)a. \tag{8.2}$$

acts on the compensation plate K_4.

This yields on the electrodes E_2 and E_4 a resulting electric charge of

$$Q = 3d_{11}(m_M + m_D)a \tag{8.3}$$

and on the electrode E_1 the electric charge of

$$Q^* = -d_{11}^*(m_M + m_D + m_S)a. \tag{8.4}$$

For a complete acceleration compensation, both electric charges must cancel each other, i.e.

$$Q + Q^* = 3d_{11}(m_M + m_D)a - d_{11}^*(m_M + m_D + m_S)a = 0. \tag{8.5}$$

If the sensitivities of all piezoelectric elements are the same, i.e. $d_{11} = d_{11}^*$, it follows

$$3(m_M + m_D) = m_M + m_D + m_S \tag{8.6}$$

and

$$m_S = 2(m_M + m_D). \tag{8.7}$$

When measuring pressure, all plates are subjected to the same force F_D and the resulting electric charge is

$$Q_D = (3d_{11} - d_{11}^*)F_D = 2d_{11}F_D. \tag{8.8}$$

The actual piezoelectric sensitivity to pressure corresponds to that of two such piezoelectric plates.

By choosing appropriate materials or crystal cuts, it is possible to achieve that $d_{11} > d_{11}^*$. Equation (10.7) then becomes

$$m_S = \frac{3d_{11} - d_{11}^*}{d_{11}^*}(m_M + m_D). \tag{8.9}$$

Another type of acceleration-compensated pressure sensor is shown in Fig. 8.14. This time, the main transduction element consists of quartz elements cut for the transverse effect, having a cross sectional area of A_S, an electrode with an area of A_E and the piezoelectric coefficient d_{12}. As compensating piezoelectric element, a quartz plate cut for the longitudinal effect with d_{11} is used. For m_S, we obtain

$$m_S = \frac{\frac{d_{12} A_E}{A_S} - d_{11}}{d_{11}} \left(m_M + m_D\right). \tag{8.10}$$

For an even better acceleration compensation, the inertial influence of other parts such as the piezoelectric elements themselves, the electrodes and so on must be taken into account. In practice, the final design is best found by empirically finalizing the exact dimensions of the various parts.

The method of acceleration compensation just described is also known as "electric elimination of inertial forces" and was first described by Gohlke [1955, p 101], albeit with some errors in the formulae derived. Attempts were made also to obtain an acceleration-compensated sensor by "suspending" the piezoelectric elements on springs [Gohlke 1955, p 104]. This proved to be unsuccessful in practice, mainly because it reduced the natural frequency to an unacceptably low value.

A novel approach of acceleration compensation is passive compensation by inertia [Bill et al 2001; Engeler 2000]. The principle is represented in Fig. 8.15. Instead of a separate seismic mass and an additional transduction element as shown in Fig. 8.14, the peloading sleeve is attached at the front and the part behind the transduction elements forms the seismic mass m_s. Denoting the mass of the diaphragm with m_d, the spring constant of the diaphragm with k_d and that of the

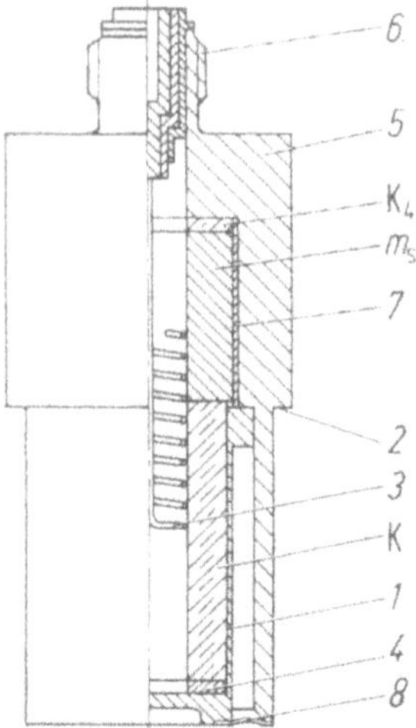

Figure 8.14 Acceleration-compensated pressure sensor with quartz elements for the transverse effect (Courtesy of Kistler). *1* preloading sleeve, *2* sealing shoulder, *3* contact spring for connecting the vacuum-deposited electrodes on the quartz elements for the transverse effect with the seismic mass m_s acting as electrode for the X-quartz plate K_4 and with the connector *6*, *4* pressure transmitting piece, *5* housing, *6* connector, *7* insulation, *8* diaphragm, K quartz elements for the transverse effect, K_4 quartz element for the longitudinal effect

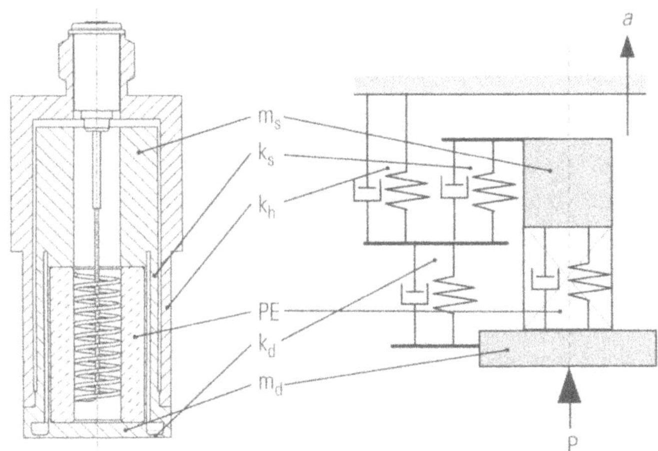

Figure 8.15 Passive acceleration compensation (Courtesy of Kistler) m_s seismic mass, k_s spring constant of seismic mass support (is also preloading sleeve for transduction element), k_h spring constant of housing (has no influence on acceleration compensation, but on natural frequency of sensor), PE piezoelectric element, k_d spring constant of diaphragm, m_d mass of diaphragm, p acting pressure (measurand), *a* acceleration

preloading sleeve with k_s, we can write for small damping values δ: $m_d \cdot a/k_d = m_s \cdot a/k_s$, i.e. $m_d/k_d = m_s/k_s$. As the mass m_d of the diaphragm and the spring constants are given, the seismic mass m_s must be adjusted to fulfill the condition for acceleration compensation. In practice this is done experimentally.

8.7 Pressure Sensors for High Temperature

Piezoelectric pressure sensors usually have an upper operating temperature limit of about 200 to 250 °C. Pressure in fluids of a higher temperature can be measured with these sensors if water cooling can be used. Water cooling has several drawbacks: pressure sensors mounted in water-cooled adapters require more space for mounting, their frequency response may be impaired at the upper end and the flow of the cooling water may cause noise when measuring low pressure. Directly cooled pressure sensors usually exhibit a higher acceleration error. Moreover it is often just not possible to install water cooling because of the required tubing, pump and reservoir.

Pressure sensors of special design have been developed, allowing uncooled operation over 350 °C. There are two major problems to be solved: a piezoelectric material with a sufficiently wide operating temperature range must be found for the transduction element, and insulation materials to match that temperature range. PTFE can be used barely up to 250 °C before it starts to soften and disintegrate. For temperatures above 250 °C practically only ceramic materials can be used as insulator.

Several piezoelectric materials can be used at higher temperature (see also chapter 3). Quartz is usable up to close its transition temperature of 573 °C where it changes from the α-phase into the β-phase (also piezoelectric but with completely different coefficients). The piezoelectric coefficients of quartz drop with increasing temperature and quartz is prone to twinning under high mechanical load (see 3.2.4). Special quartz cuts have been found such as the Polystable® cut which are highly resistant against twinning and whose piezoelectric sensitivity changes little up to over 400 °C. Crystals of the CGG group (see 3.5) can be used at much higher temperatures, i.e. up to around 600 °C. The Kistler company has succeeded in growing such crystals and is using them in sensors for cylinder pressure in internal combustion engines and cavity pressure in injection molding of plastics as well as in force and strain sensors [Gossweiler and Cavalloni 1999]. Gallium orthophosphate (see 3.4) is a viable alternative too, for temperatures up to about 600 °C.

Tourmaline has been used for a long time in high-temperature sensors, because it works up to about 650°C. However, it exhibits a strong pyroelectric effect and its insulation resistance drops considerably at higher temperature, allowing only dynamic measuring above a lower frequency limit of about 10 Hz. Only a few pressure sensors with tourmaline elements have been built, mainly for special applications in the field of nuclear reactors. Other high-temperature applications are:

- measuring pressure fluctuations in e.g. the combustion chambers of gas turbines to detect flame instabilities,
- detecting jet pipe resonance in the jet pipe of the afterburners in military jet engines,
- rotating swirl detection in high-pressure compressors of jet engines.

For sensors working at such high temperatures, steel is not a suitable material. Instead, nickel-based superalloys must be used.

Piezoelectric ceramics have similar limitations, i.e. a strong pyroelectric effect and a low insulation resistance, allowing only dynamic measuring above about 10 to 50 Hz at higher temperature. For this reason, piezoelectric ceramics are rarely used in pressure sensors.

8.8 Sensors for Cylinder Pressure in Internal Combustion Engines

One of the first applications of piezoelectric pressure sensors – measuring the cylinder pressure in internal combustion engines – still is of prime importance in the field of engine research. The first known publications are from Japan [Okochi et al 1925] and go back to 1925, when Okochi designed the first quartz pressure sensor with 2 quartz plates cut for the longitudinal effect as transduction element. Since then quartz sensors have been the preferred choice for measuring cylinder pressure [Kuratle 1995]. A number of types have been developed to meet the sometimes contradicting requirements such as resistance to knocking, measuring the pressure with an error of less than 1% if not 0,1% even under changing temperature and having a life time of over 10^7 pressure cycles.

Figure 8.16 High-temperature, ground-isolated pressure sensor for continuous operating in combustion engines up to 350 °C (Courtesy of Kistler)

Sensors used for cylinder pressure in combustion engines are subjected to high flash temperatures, reaching intermittently over 2500 °C. A number of diaphragm designs (e.g. with an incorporated ceramic plate acting as a thermal buffer) and special TiN coating to obtain a long life have been developed (Fig. 8.16). Such diaphragms have a certain mass which limits the natural frequency of the sensor to around 80 kHz.

For precise indication work, water-cooled sensors whose temperature can be kept within close tolerance are still required. Although the temperature of the transduction element is kept by proper water cooling at around 50 °C, it is advisable to use sensors that support much higher temperature to prevent damage

Figure 8.17 High-temperature sensor for cylinder pressure in combustion engines. A crystal from the CGG group (see 3.5) is used as transduction element which gives the high sensitivity of 19 pC/bar and a continuous operating range up to over 350 °C. Natural frequency is ≈130 kHz (Courtesy of Kistler)

Figure 8.18 High-temperature sensor for cylinder pressure in combustion engines. A transduction element made of gallium orthophosphate (see 3.4) is used, resulting in a sensitivity of 15 pC/bar and a continuous operating range up to 400 °C. Natural frequency is ≈115 kHz (Courtesy of AVL)

should the water cooling fail. Sensors with Polystable® quartz elements offer such a protection without compromising the accuracy of measuring.

With the introduction of new crystals such as gallium orthophosphate and crystals from the CGG group, exhibiting similar characteristic as quartz but offering a much wider temperature range, it has become possible to measure precisely even without water cooling.

Another critical area is the inherent correlation between sensitivity and diameter of the the sensor front. Combustion engines, especially the multi-valve type, offer very little space for mounting a pressure sensor with its front close to or even flush with the combustion chamber wall. Decreasing the front diameter reduces the active area of the diaphragm and correspondingly the resulting force on the transduction element originating from the same applied pressure. Sensor with sensitivities less than about 10 pC/bar have an insufficient signal-to-noise ratio.

The new crystals from the CGG group and gallium orthophosphate offer almost triple the sensitivity of quartz and sensors with such transduction elements are already on the market (Fig. 8.17 and 8.18).

8.9 Sensors for Cavity Pressure in Plastics Processing and in Die Casting of Metals

The course of the pressure in the cavity of the mold during injection molding of plastics as well as in die casting of metals is the key parameter for the control of the process. The cavity pressure can be measured in two ways: *directly* by a sensor mounted flush with the wall of the cavity or *indirectly* by a force sensor placed behind a special pin of the same design as a standard ejector pin.

Early attempts to measure cavity pressure directly failed for several reasons. Sensors with a diaphragm – which has to be quite thin for obtaining sufficient sensitivity – will not live long enough because the melt tends to stick and eventually tear the diaphragm away. Also such diaphragms are likely to leave a mark on the molded part, which is rarely acceptable. Moreover sensors with diaphragm can only be mounted in a flat spot of the cavity wall because their front can not be machined to match the contour or structure of the wall.

For measuring the cavity pressure indirectly, quartz force sensors (Fig. 6.8) and load washers (Fig. 6.2) have been used for a long time. Their small size and high rigidity allow them to be mounted behind an ejector pin (this can be one of the regular ejection pins of a mold or a separate pin used only for measuring) whose front is flush with the wall of the cavity in the mold. While this solution has the advantage that the sensor is not directly exposed to the hot melt (usually around 250 to over 400 °C, depending on the material) there are serious drawbacks. The pin which transmits the pressure to the sensor may slightly bend under the influence of the temperature which leads to additional friction and stick-slip in the close-tolerance bores which hold the pin. Although the axial movement of the pin is in the order of a few μm only, hysteresis and lower repeatability result.

Measuring the cavity pressure directly became possible with the first quartz sensors designed specifically for this applications, introduced in the late 70's. These cavity pressure sensors do not have a diaphragm but work on the same principle as ejector pins: The sensor has a cylindrical front in whose center is mounted a force-sensitive pin, transmitting the melt pressure directly to the piezoelectric transduction element (Fig. 8.19). The annular gap between pin and surrounding cylinder is – as with an ejector pin – only a few μm wide and the melt will not penetrate because of its viscosity (Fig. 8.20). The front is of solid steel and can easily be machined to match, within limits, the inclination, the curvature and the surface structure of the cavity wall. The life time of such sensors is practically unlimited. The massive steel pin of these sensors offers also the advantage that it will not – unlike a thin diaphragm – warp during temperature transients which would cause a change in preload on the transduction elements, causing a large temperature gradient error.

For use in small molds, sensors without external steel cylinder (part *1* in Fig. 8.21) are used. They can be fitted with an O-ring seal in a mounting bore of close tolerance and also be machined to match the cavity wall. The miniature type with 2,5 mm front diameter is successfully used in molding engineering plastics.

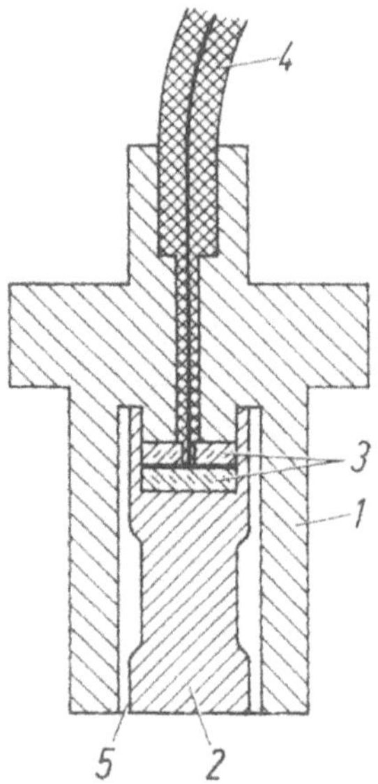

Figure 8.19 Cavity pressure sensor for plastics processing (after Kistler). *1* sensor housing, *2* steel cylinder, *3* transduction element, *4* cable, *5* annular gap of a few μm

Figure 8.20 Cavity pressure sensor of the design shown in Fig. 8.18 with a front diameter of 8 mm (Courtesy of Kistler)

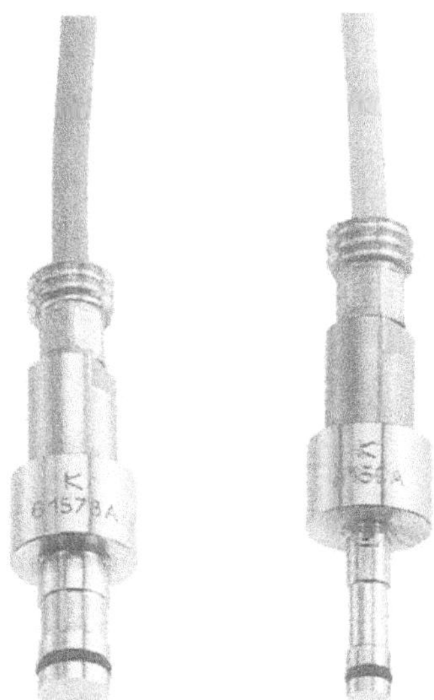

Figure 8.21 Cavity pressure sensors without external cylinder and direct O-ring sealing in the mounting bore, with front diameters of 5 mm and 2,5 mm (Courtesy of Kistler)

Sensors for measuring the cavity pressure in die casting of metals (aluminum and zinc) are of similar construction, but have a front designed to withstand the considerably higher melt temperatures encountered in these metals.

Cavity pressure sensors use quartz elements of a special cut which makes them resistant against twinning and work with melt temperatures over 400 °C. Their measuring range goes up to 2 kbar, which covers practically all applications in plastics injection molding.

These sensors without diaphragm must never be calibrated with a gaseous or low-viscosity fluid, because it would penetrate into the annular gap and exert a radial pressure on the thin preloading sleeve around the transduction elements. This would falsify the calibration and most likely even destroy the sensor. These sensors may be calibrated by applying a force directly on the front of the central pin whose diameter and therefore area is precisely known. Or it suffices to simply create a temporary "diaphragm" by applying a piece of ordinary adhesive tape (e.g. 3M Magic Mending Tape®) and calibrating with a fluid (e.g. oil).

8.10 Calibration of Pressure Sensors

Pressure sensors are calibrated by applying precisely known pressures through a fluid on their diaphragm and recording the corresponding output signals.

The most precise source for pressure is the so-called "dead weight tester", also called "piston gage" or "piston manometer" (Fig. 8.22). A piston made of steel (for high pressure also made of tungsten carbide having a higher Youngs' modulus),

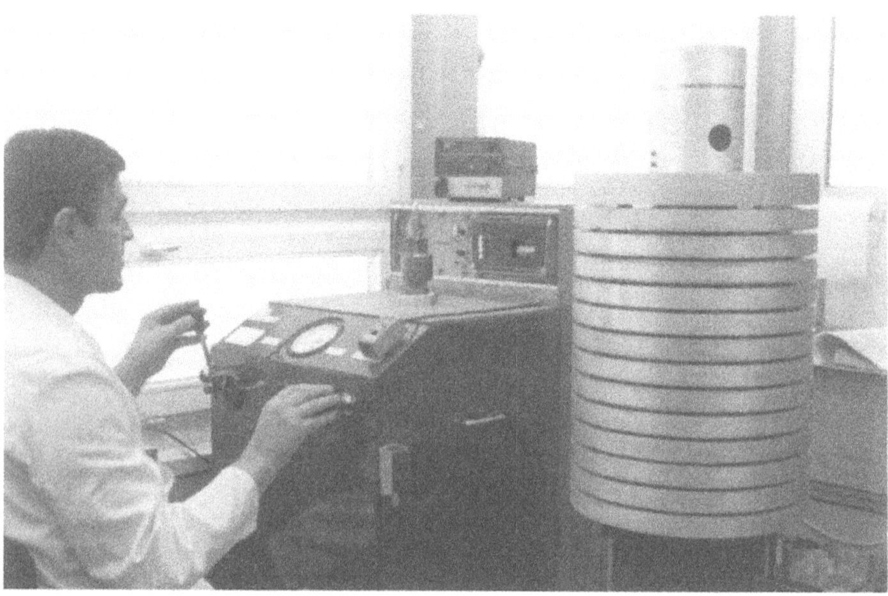

Figure 8.22 Dead weight tester for calibrating pressure sensors up to over 8 kbar (Courtesy of Kistler)

machined to close tolerance, is held in a matching vertical cylinder and can be loaded by adding exactly known weights in the form of discs. The pressure of the oil acting from below on the piston and also on the sensor to be calibrated can be increased either by a hand-operated spindle or by a pump. When the piston starts to float, the pressure has reached a value $p = F/A$, with $F = m \cdot g$ (m is the total mass of the piston and the added weights) and A the cross-sectional area of the piston. Since F, m and g can easily be determined with great precision, pressure so generated has an error of less than 0,05%. An important detail is to let the piston rotate when the floating state is reached in order to eliminate any influence of friction. Usually a small electric motor is built into the dead-weight tester to that end.

The sensor to be calibrated (it is possible to calibrate several sensors at the same time) can now be exposed to the various pressures and the output recorded. Although this method is very precise, it has several practical drawbacks. The pressure can only be increased in discrete steps (corresponding to the weight that is added) and not continuously. While increasing the pressure is rather easy to handle, reducing the pressure requires great skill by the operator. After lifting off a weight great care must be exercised to prevent the pressure from dropping too low, requiring a certain increase to reach the desired value again, i.e. the piston must drop into the floating position without first hitting the lower stop. Otherwise the reading is falsified by the hysteresis! Recording a correct calibration curve, be it in a continuous manner or in discrete steps, demands that first the pressure is always increasing until the maximum is reached and then always decreasing until zero is

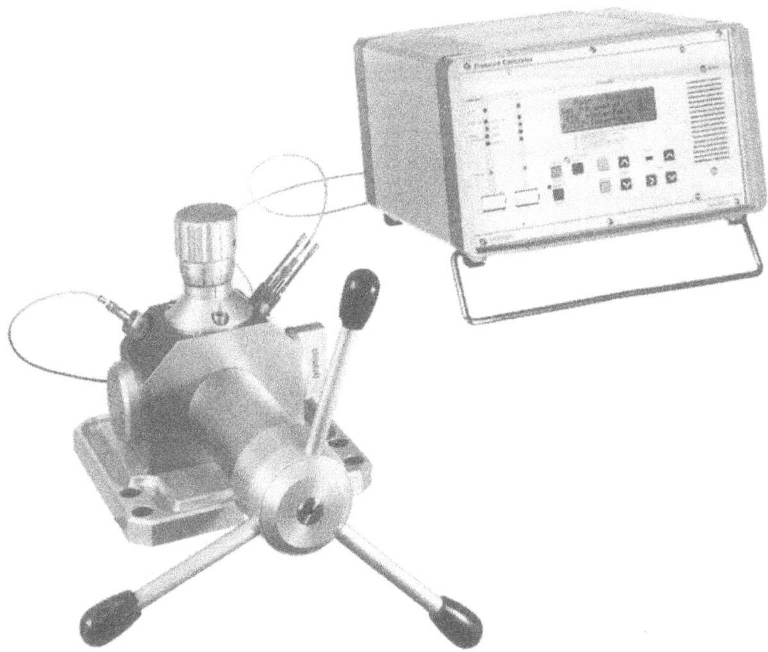

Figure 8.23 Pressure calibration system consisting of a high-pressure generator (0...7kbar) and a calibrator (Courtesy of Kistler)

reached again. Only then can the hysteresis be determined correctly. Obviously these procedures are rather time-consuming and therefore are usually only used when the highest accuracy is required.

For routine calibration, a sensor calibrated on a dead-weight tester is used as working standard (a reference sensor with traceably calibration). The sensor or group of sensors to be calibrated are installed in the same pressure chamber of a calibration device and the pressure is then generated by a spindle-driven pressure piston. The outputs from both are recorded simultaneously and the sensitivity, linearity and hysteresis can then easily be calculated. With proper care, the calibration error (or uncertainty) can easily be kept below about 0,3%, provided the charge amplifiers and recording equipment used are calibrated electrically to within the needed tolerance.

Some manufacturers offer special pressure calibrator systems consisting of reference sensors and amplifiers with traceable calibration, and suitable pressure generators. Fig. 8.23 illustrates such a calibration system for pressure up to 8kbar. The sensor to be calibrated is compared with a reference sensor of traceable calibration. Systems for calibrating in the 10kbar range are described e.g. in [Jäger et al 1996].

Another approach are hydraulic impulse generators. For high-pressure calibration, a system with a free falling known mass can be used (Fig. 8.24).

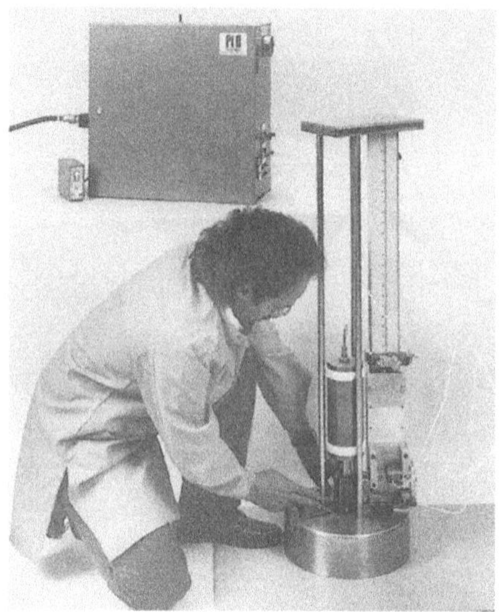

Figure 8.24 High-pressure impulse generator for calibrating pressure sensors. The system produces pressure pulses of about 7 ms width, with a rise time of about 3,5 ms and maximum pressure peaks up to 8,6 kbar (Courtesy of PCB)

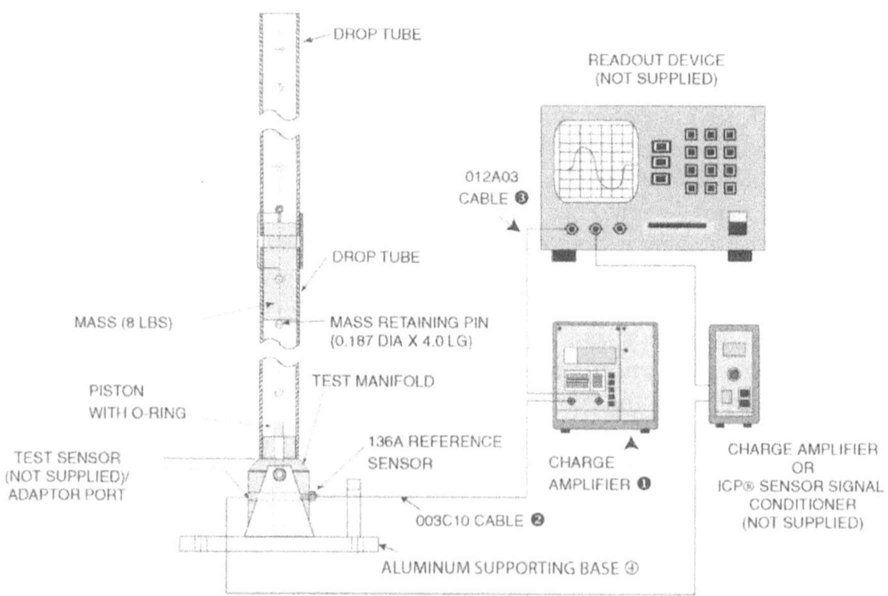

Figure 8.25 Pressure impulse calibration system. The system produces pressure pulses of about 5 ms width, with a rise time of about 2 ms and maximum pressure peaks of to 1,4 kbar (Courtesy of PCB)

Figure 8.26 Reference pressure sensor used in the calibration system shown in Figs. 8.25 and 8.25. The epoxy-sealed tourmaline transduction element exploits the hydrostatic piezoelectric effect. Range: 0…1,4 kbar, natural frequency: 1 MHz, rise time: 3 μs (Courtesy of PCB)

A precision accelerometer attached to the falling mass measures the deceleration when the mass hits the piston of the pressure chamber. From the measured deceleration, the known mass and geometry of the pressure chamber, the pressure can be deduced. The system is claimed to measure within about ±0,3 % error. Still another approach is a hydraulic pressure calibrator in whose pressure chamber a special tourmaline sensor is mounted (Figs. 8.25). Tourmaline exhibits a hydrostatic piezoelectric effect (see 3.3), which makes it very suitable for measuring pressure directly immersed in oil inside a pressure chamber (Fig. 8.26). A dropping mass produces impulses up to about 1,4 kbar with a pulse width of 5…6 ms and a rise time of about 2…3 ms.

9 Acceleration Sensors

9.1 Quantity and Units of Measurement

The quantity "acceleration" is defined as the time rate of change in velocity and represented by the symbol a. The unit for acceleration in the SI (Système International) is m/s^2. In general engineering practice, especially for measuring vibration, often the acceleration of free fall g is preferred as a working unit. Because the acceleration of free fall (also called "acceleration due to gravity") slightly varies with the geographical location, the "standard acceleration of free fall", defined as g_n=9,80665 m/s^2 (exactly), is generally used. The working unit g must not be combined with decimal prefixes.

The quantity "acceleration" is often subdivided into three types of acceleration, each of distinctly different nature: *acceleration*, *shock* and *vibration*. An overview of these fields can be found in [Harris 1996; ISO 1995; McConnell 1995; Plankey and Wilson 2001].

9.2 Basic Properties of Acceleration Sensors

Acceleration sensors are basically force sensors with an attached mass. The mechanical quantity measured is a force which is proportional to the acceleration, according to Newton's second law of motion $F=m \cdot a$. The mass m is constant and usually called "seismic mass", a term taken from seismometers used for detecting earth quake tremors. The inertia of the seismic mass m produces a force F when being accelerated and this force is measured by the acceleration sensor. The relation between a and F is determined once and for all when calibrating the sensor.

The basic characteristics of an accelerometer are briefly analyzed in a simple model. We chose a compression-type acceleration sensor with central preload, but the results are analogously valid for other types of design, too. The elementary design is illustrated in Fig. 9.1. The transduction element consists of 2 piezoelectric plates cut for the longitudinal effect (*1*) and oriented with their polarities in acts

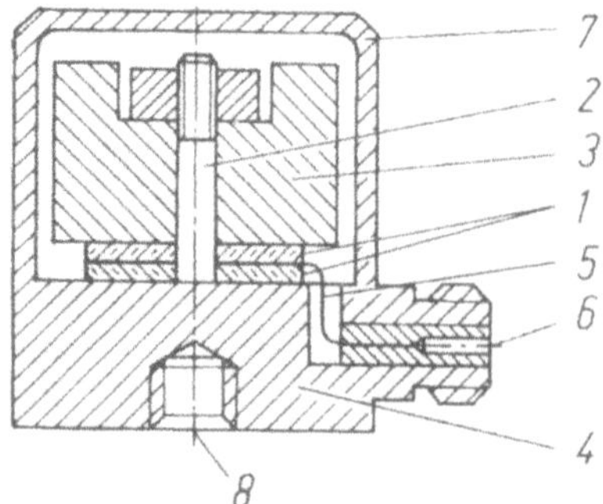

Figure 9.1 Design of a simple acceleration sensor

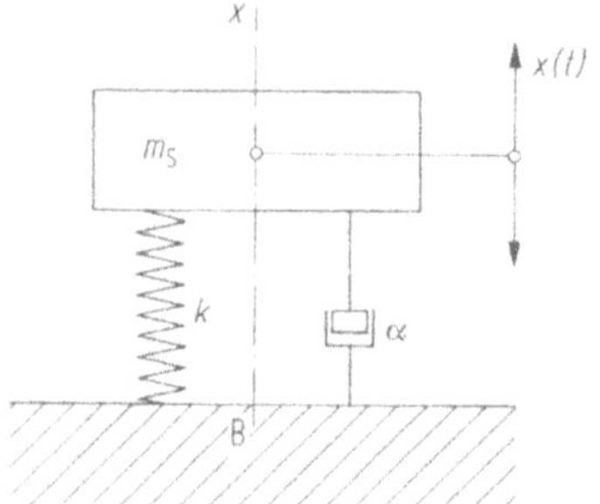

Figure 9.2 Model of an oscillating system with 1 degree of freedom

opposition against the central electrode. They are preloaded by a bolt (*2*) which as a very rigid spring between the seismic mass (*3*) and the base plate (*4*). The electrode (*5*) captures the output signal and feeds it to the connector (*6*), The housing (*7*) protects the sensor from the environment and also has the mounting thread (*8*). When the base plate is accelerated, the seismic mass exerts a proportional force on the transduction element.

Such a sensor is a oscillating system with one degree of freedom, represented by a very simplified model in Fig. 9.2. It consists of a seismic mass m_s, a spring with a spring constant k (effective rigidity) and a damping element with a damping constant α. The spring attaches the seismic mass on the base B. We assume that the system can only oscillate in one direction, called x-axis. The coordinate system is attached to the base and its origin is chosen as the position of the center of gravity of the seismic mass when the system is at rest. If the system has linear and not time-variant properties and there is no external force acting, the motion of the seismic mass can be described by the equation of motion

$$m_S \frac{d^2x}{dt^2} + \alpha \frac{dx}{dt} + kx = 0. \tag{9.1}$$

If an external exciting force $F(t)$ acts on the system, the right side can be extended by

$$m_S \frac{d^2x}{dt^2} + \alpha \frac{dx}{dt} + kx = F(t). \tag{9.2}$$

The solution of this equation yields the displacement $x(t)$ of the center of gravity of the seismic mass, which was provoked by the excitation $F(t)$. The velocity $v(t)$ *or the acceleration* $a(t)$ of the seismic mass can be derived from the definitions

$$v = \frac{dx}{dt}, \tag{9.3}$$

respectively

$$a = \frac{d^2x}{dt^2}. \tag{9.4}$$

The time-dependence of $F(t)$ can be anything. The principle of superposition, which states that the effect of simultaneous and superposed events is equal to the sum of the effects of the individual events, can be applied in solving (11.2).

If we neglect the damping in (11.1) and set $\alpha = 0$, we have a harmonic motion, and the general solution is

$$x = A \sin(\omega_0 t + \varphi), \tag{9.5}$$

whereby

$$\omega_0 = \sqrt{\frac{k}{m_S}} \tag{9.6}$$

represents the frequency of the system. It depends on the period and the natural frequency by the known relationships

$$\omega_0 = 2\pi f_0 \tag{9.7}$$

and

$$\omega_0 = \frac{2\pi}{T_0}. \tag{9.8}$$

The time-independent amplitude A and the phase shift φ are obtained from the initial conditions.

If the damping is not ignored in (9.1), the solution depends on the damping which follows from considering the corresponding characteristic equation.

The critical damping (critical damping constant) is defined as

$$\alpha_K = 2m_S\sqrt{\frac{k}{m_S}} = 2m_S\omega_0 \tag{9.9}$$

and determines the dimensionless degree of damping (damping ratio) θ through the ratio of the considered damping to the critical damping

$$\vartheta = \frac{\alpha}{\alpha_K} = \frac{\alpha}{2m_S\omega_0}. \tag{9.10}$$

For very small damping ($\theta \ll 1$) which is significant in acceleration sensors, the general solution of (9.1) can be written in the form

$$x = A\ \mathrm{e}^{-\frac{\alpha t}{2m_S}} \sin(\omega t + \varphi) = A\ \mathrm{e}^{-\delta t} \sin(\omega t + \varphi). \tag{9.11}$$

Once such a system is disturbed from its equilibrium position and then left to itself, it will oscillate with decreasing amplitude (Fig. 9.3) at the natural angular frequency

$$\omega_d = \sqrt{\frac{k}{m_S}\left(1-\vartheta^2\right)} = \omega_0\sqrt{1-\vartheta^2} \tag{9.12}$$

which is lower than the natural angular frequency of a similar but undamped system ($\omega < \omega_0$). In (9.11) stands for the fading constant δ

$$\delta = \frac{\alpha}{2m_S}. \tag{9.13}$$

It can easily be shown that the ratio of 2 consecutive amplitudes A_n and A_{n+1} of the same polarity is constant in a damped oscillation, i.e.

$$\frac{A_n}{A_{n+1}} = \mathrm{e}^{\frac{\pi\alpha}{m_S\omega}} = \text{const.} \tag{9.14}$$

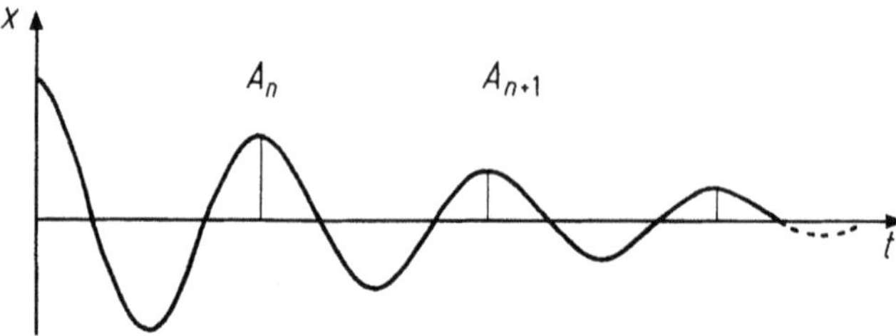

Figure 9.3 Damped oscillation

The natural logarithm of this ratio called the logarithmic decrement Δ

$$\Delta = \ln\left(\frac{A_n}{A_{n+1}}\right) = \frac{\pi\alpha}{m_S\omega} = \frac{2\pi\vartheta}{\sqrt{1-\vartheta^2}}. \tag{9.15}$$

For critical damping (for $\alpha = \alpha_k \Rightarrow \theta = 1$), the general solution is

$$x = (A + Bt)e^{-\frac{\alpha t}{2m_S}} \tag{9.16}$$

and with a still stronger damping (for $\alpha > \alpha_k \Rightarrow \theta > 1$)

$$x = e^{-\frac{\alpha t}{2m_S}}\left[A\,e^{\omega t} + B\,e^{-\omega t}\right], \tag{9.17}$$

with

$$\omega = \sqrt{\frac{k}{m_S}\left(\vartheta^2 - 1\right)}. \tag{9.18}$$

No oscillation develops and the seismic mass returns aperiodically to its equilibrium position (Fig. 9.4).

The solution of (9.2) depends on the time-dependence of the exciting force $F(t)$. If it is sinusoidal, i.e.

$$F(t) = F_0 \sin\omega t, \tag{9.19}$$

the general solution, known from the basic theory of oscillations, in the stationary condition (after the oscillation has settled in) is

$$x = A\sin(\omega t - \varphi), \tag{9.20}$$

whereby the amplitude is frequency-dependent, i.e.

$$A = \frac{F_0}{\sqrt{\left(k - m_S\omega^2\right)^2 + (\alpha\omega)^2}}, \tag{9.21}$$

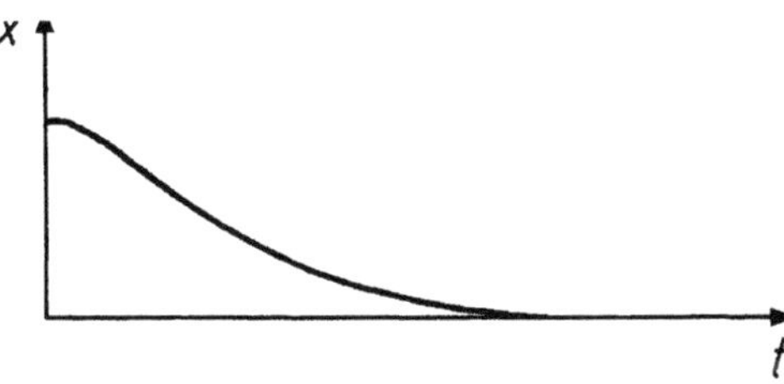

Figure 9.4 Over-critically damped oscillation

and the phase of the oscillation lags behind the phase of the excitation by the value of φ, which works out as

$$\tan\varphi = \frac{\alpha\omega}{k - m_S\omega^2}. \tag{9.22}$$

For $\omega = 1$, the amplitude – which equals the displacement caused by the static force F_0 – is obtained from (9.21) as

$$A_0 = \frac{F_0}{k}. \tag{9.23}$$

Considering (11.6), we can convert (9.21) and (9.22) into a dimensionless form by writing

$$\frac{A}{A_0} = \frac{1}{\sqrt{\left(1 - \frac{\omega^2}{\omega_0^2}\right)^2 + \left(2\vartheta\frac{\omega}{\omega_0}\right)^2}} \tag{9.24}$$

and

$$\tan\varphi = \frac{2\vartheta\frac{\omega}{\omega_0}}{1 - \left(\frac{\omega}{\omega_0}\right)^2}. \tag{9.25}$$

The dependence of the amplitude ratio A/A_0 and of the phase angle φ on the frequency ratio, with the damping ratio δ as parameter, is shown in Fig. 9.5. By differentiating and setting to zero the denominator in (9.24) we obtain

$$\omega_{max} = \omega_0\sqrt{1 - 2\vartheta^2}. \tag{9.26}$$

The peak values of the curves in Fig. 9.5 move towards lower frequencies when damping is increasing. For a low damping the frequencies ω_0, ω_d and ω_{max} are nearly equal. ω_{max} is called the resonant frequency of the externally excited damped oscillations. An acceleration sensor can not be used near the resonant frequency because there both, amplitude and phase, depend heavily on the frequency.

However it is possible to measure with an acceleration sensor in a frequency range far above its natural frequency. Now the seismic mass remains at rest and the output signal is proportional to the displacement of the measured object (oscillation displacement sensor, also called "vibrograph"). Due to their extremely high rigidity and inherent high natural frequency, piezoelectric sensors are not suitable for such applications.

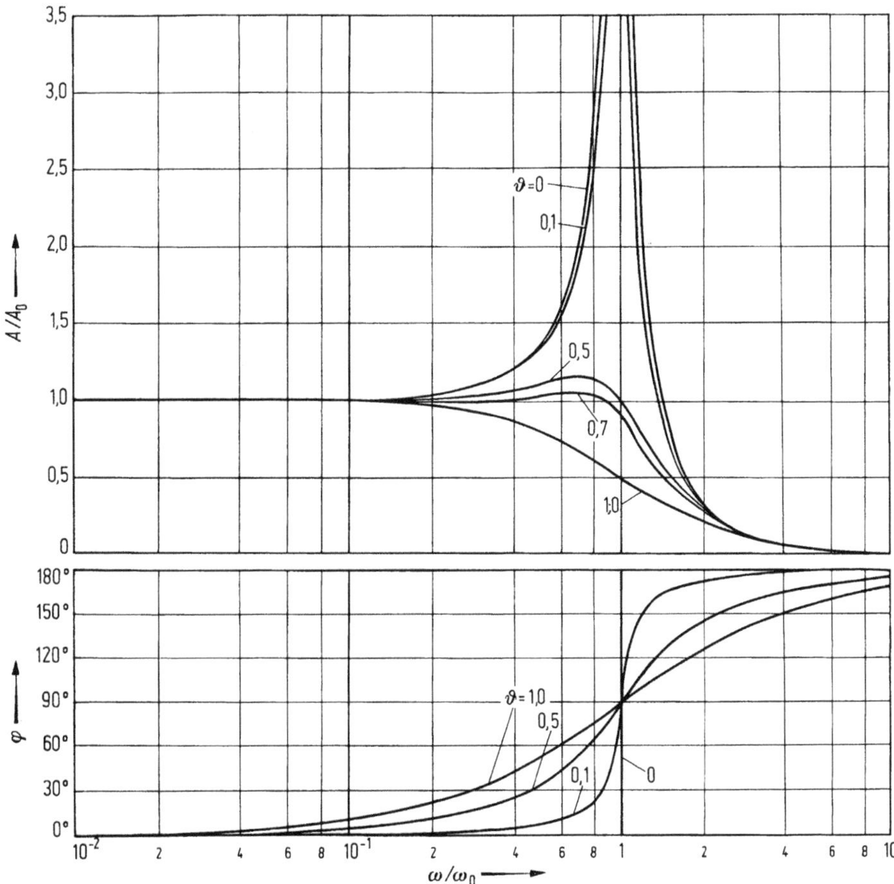

Figure 9.5 Dependence of the amplitude ratio A/A_0 and the phase angle ϕ on the frequency ratio ω/ω_0 for different damping factors θ

In all considerations so far we have assumed that the base (bottom) of the acceleration sensor is attached to the measured object in an ideally rigid manner and that the behavior of the measured object is not altered by the attached sensor. In practice this is never true. For analyzing the influence of the mechanical connection between sensor and object, we replace the model of Fig. 9.2 by the model in Fig. 9.6.

We assume the measured object to oscillate sinusoidally , i.e.

$$x_U = A_U \sin\omega t. \tag{9.27}$$

For the masses m_S and m_B, the following equations of motion apply:

$$m_S \frac{d^2 x_S}{dt^2} + k\left(x_S - x_B\right) = 0. \tag{9.28}$$

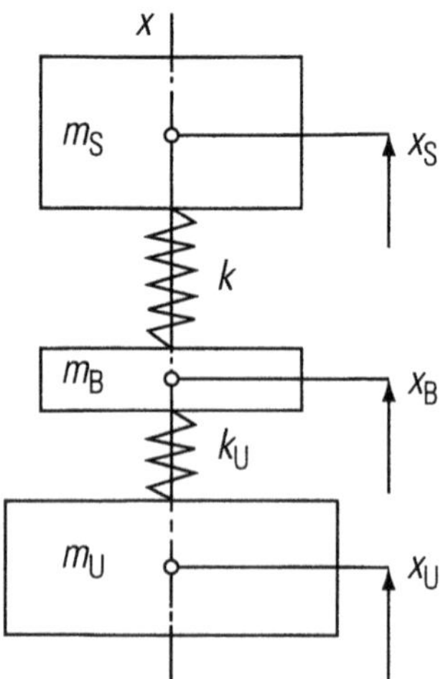

Figure 9.6 Model of an acceleration sensor and its coupling to the measured object

and

$$m_B \frac{d^2 x_B}{dt^2} + k(x_B - x_S) + k_U(x_B - x_U) = 0. \tag{9.29}$$

At resonance the elongation x_S x_B of the seismic mass relative to the base becomes maximal. When approaching the resonant frequency from below all 3 masses oscillate at the same phase as the measured object and

$$x_S = A_S \sin \omega t \tag{9.30}$$

and

$$x_B = A_B \sin \omega t. \tag{9.31}$$

apply.

After inserting in (9.28) and (9.29) we obtain

$$\begin{aligned} &\left(k - m_S \omega^2\right) A_S - k A_B = 0 \\ &-k A_S + \left(k + k_U - m_B \omega^2\right) A_B = k_U A_U. \end{aligned} \tag{9.32}$$

At resonance, A_S A_B as well as A_S and A_B each must reach a maximum value, too. This is so when the determinant of the coefficients of A_S and A_B at the left side of the system of equations (9.32) disappears (it will be in the denominator of the solution for A_S and A_B). For ω we obtain a bi-quadratic equation

$$\omega^4 - \left[k\left(\frac{1}{m_S}+\frac{1}{m_B}\right)+k_U\frac{1}{m_B}\right]\omega^2+\frac{kk_U}{m_S m_B}=0. \tag{9.33}$$

We introduce the dimensionless parameters

$$a=\frac{m_B}{m_S},\ r=\frac{k_U}{k},\ \omega_0=\sqrt{\frac{k}{m_S}}. \tag{9.34}$$

and write

$$\left(\frac{\omega^2}{\omega_0^2}\right)^2-\left(1+\frac{1}{a}+\frac{r}{a}\right)\frac{\omega^2}{\omega_0^2}+\frac{r}{a}=0. \tag{9.35}$$

In solving the equation we look at the lowest resonant frequency which is of practical interest and obtain

$$\frac{\omega^2}{\omega_0^2}=\frac{\left[1+\frac{1}{a}(1+r)\right]-\sqrt{\left[1+\frac{1}{a}(1+r)\right]^2-\frac{4r}{a}}}{2}. \tag{9.36}$$

Of course this solution includes the previously discussed ideal case, too. If the base is attached rigidly to a very large (infinite) mass of the measured object, r and a will tend to infinity ($r\to\infty\wedge a\to\infty$) and we obtain the same limit value as in (9.6)

$$\omega=\omega_0=\sqrt{\frac{k}{m_S}}. \tag{9.37}$$

The only parameters determining the natural frequency are m_s and k. The not ideal coupling between the base and the measured object reduces the resonant frequency as illustrated in Fig. 9.7, which shows the dependence of ω/ω_0 on the ratio of the two spring constants $r=k_U/k$ with the ratio $a=m_B/m_S$ of the mass of the base to the seismic mass as the parameter. Fig. 9.8 gives the frequency response curves of an acceleration sensor mounted in different ways.

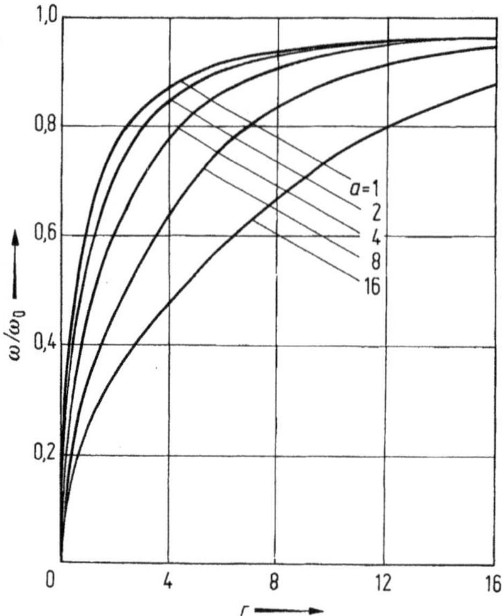

Figure 9.7 Influence of the nonideal coupling between acceleration sensor and measured object

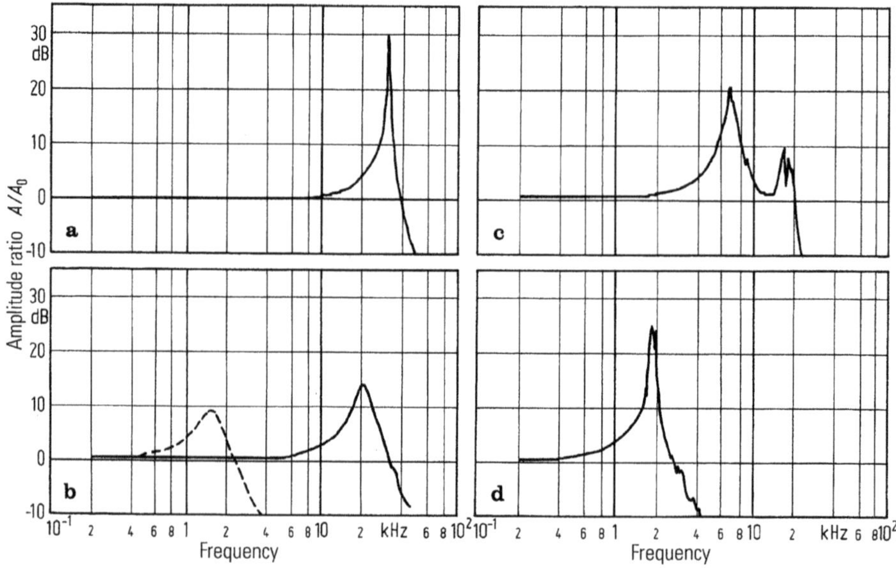

Figure 9.8 Effects of various ways of attaching one and the same acceleration sensor to the measured object (after Brüel & Kjær). **a** screwed-on with a steel stud, contact faces lightly coated with silicone grease (similar results are obtained by attaching the sensor with beeswax or with cyano-acrylate adhesive, **b** attached with thin (dashed line) and thick adhesive tape, **c** sensor screwed on a magnetic base that holds magnetically on the measured object, **d** sensor screwed on a pin-shaped probe that is hand-held against the measured object

9.3 Designs of Piezoelectric Acceleration Sensors

Despite the large variety of acceleration sensors available on the market, their design can be grouped into a few basic types of construction. In principle the transduction element can only be designed to work under compression, shear or flexion.

Sensors with an element that is lo aded by compression – a force normal to its surface – are shown in Fig. 9.9. In the most simple of these so-called "compression designs" the seismic mass is pressed against this transduction element directly by the wall of the sensor housing. Although this design combines a high sensitivity with a high resonant frequency, it has several drawbacks. Because the housing serves directly as preloading element, such a sensor exhibits a high thermal and acoustic sensitivity as well. Also the tightening torque for mounting the sensor will influence the characteristics of the sensor.

These difficulties can be overcome using a separate sleeve or a central bolt for preloading so that the function of the housing is only to protect the inner parts of the sensor (Fig. 9.9, b and c, Fig. 9.10, and chapter 9.10). The mass of such sensors is a little higher, however.

All these compression designs suffer from base strain sensitivity because strain in the mounting surface on the measured object is transmitted quite directly to the transduction element, giving an error signal e.g. due to the transverse effect in the transduction element. If the base is made much more rigid by increasing its thickness the mass of the sensor is increased, too.

This disadvantage can be eliminated by the so-called "reversed" or "hanging" design, in which the transduction element is preloaded against the inner face of the top of the housing. This results in a very low base strain sensitivity and is also the preferred design for sensors serving as working standard in calibrating other acceleration sensors by the so-called "back-to-back" method (chapter 9.9 and Fig. 9.22).

Sensors incorporating transduction elements exploiting the shear effect are usually designed as shown in Fig. 9.11. Obviously the construction in Fig. 9.11a

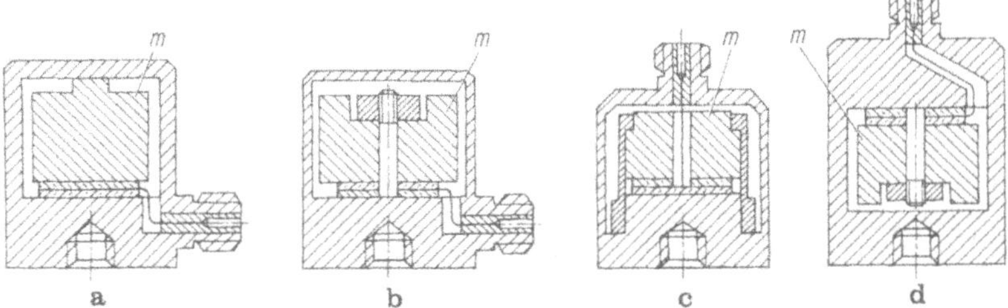

Figure 9.9 Acceleration sensors with transduction elements exploiting the longitudinal effect **a** seismic mass m is preloaded directly by the sensor housing, **b** seismic mass m is preloaded with a preloading bolt, **c** seismic mass m is preloaded with a preloading sleeve, **d** so-called "hanging design" (preloading can be by bolt or sleeve)

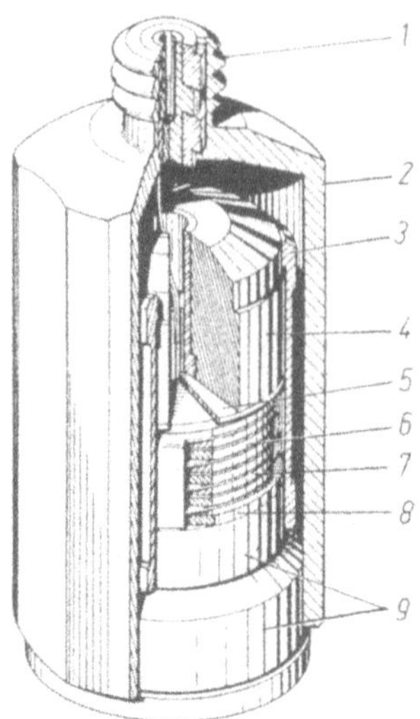

Figure 9.10 Construction of an acceleration sensor with 7 quartz plates for the longitudinal effect (Courtesy of Kistler) *1* connector, *2* housing, *3* preloading sleeve, *4* seismic mass, *5* insulating layer, *6* X-quartz plates, *7* electrodes, *8* intermediate plate for thermal compensation, *9* sensor base

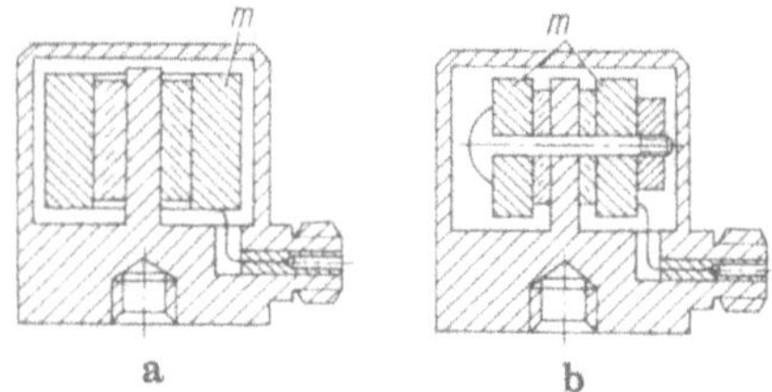

Figure 9.11 Acceleration sensors with transduction elements exploiting the shear effect **a** seismic mass m is a hollow cylinder and applies directly a radial preload on the transduction elements, **b** seismic mass m consists of two parts that are preloaded against each other by a transverse preloading bolt

can only be realized with an element made of piezoelectric ceramic because here the seismic mass is a hollow cylinder, enclosing the transduction element which is also a hollow cylinder, a shape that can not be made from quartz. The transduction element is preloaded radially against the central stud by placing the heated seismic mass on the element and producing the preload by shrinkage when cooling down. Another design uses a seismic mass in two parts, preloaded by a transverse bolt onto two transduction elements against a flat center stud [Berther et al 1997]. This design is not symmetric about the sensor axis and e.g. the cross talk ("transverse sensitivity") will not be the same in all transverse directions.

Sensors with transduction elements reacting to flexion (Fig. 9.12) have gained practical importance especially for modal analysis because they can be made at

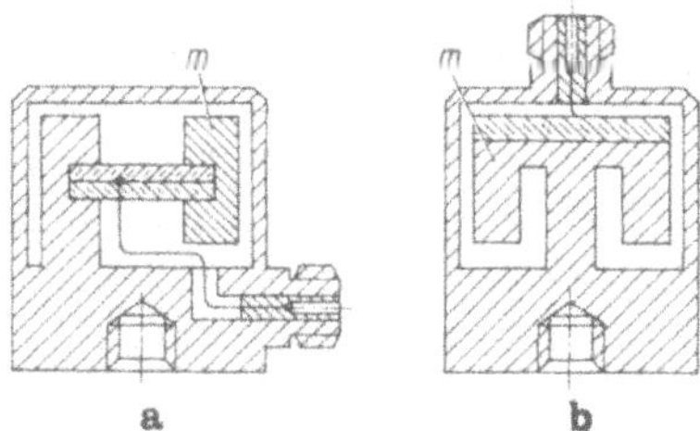

Figure 9.12 Acceleration sensors with transduction elements responding to bending **a** seismic mass m is attached to a cantilever-type transduction element, **b** seismic mass m is mushroom-shaped

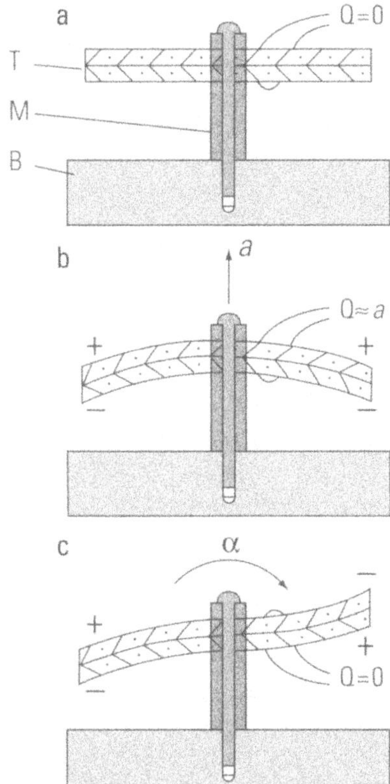

Figure 9.13 Acceleration sensor with a symmetric bending beam (double cantilever) clamped in its center (**a**). The beam **T** consists of 2 piezoelectric ceramic beams, polarized normal to their plane, and bonded together (PiezoBEAM® design by Kistler). The beam is clamped on the mounting post **M**, set on the sensor base **B**. The sensors gives an output only in response to linear acceleration a (**b**), but not to angular acceleration α (**c**)

rather low cost. A design by Kistler, called PiezoBEAM®, works according to the principle shown in Fig. 9.12 a, but without additional mass at the end of the beam. Instead, a bilaminar piezoelectric ceramic element consisting of 2 thin beams of piezoelectric ceramics is clamped in its center on top of a mounting post (symmetric double cantilever). The ceramic beams are polarized normal to their face and bonded together in a way that the electric charge yielded by the 2 beams will be added when they bend under acceleration. Such an element (Fig. 9.13) reacts only to acceleration normal to its plane but not to angular acceleration [Bill 1990]. An interesting variant of the PiezoBEAM® design is the TAP™ (Translational-Angular-PiezoBEAM®), measuring linear and angular acceleration simultaneously with the same transduction element. Instead of a single double cantilever element, 2 separate elements are clamped to the mounting post in a symmetric way (Fig. 9.14). Under linear acceleration, both elements bend in the same direction and the sum of their outputs is proportional to the acceleration normal to their plane. Angular acceleration about the y-axis will bend the elements in opposite directions and the difference of their outputs is proportional to the angular acceleration. Built-in charge amplifiers give the sensor a very high sensitivity (up to 1 V/g and 48 mV/(rad/s^2) [Bill 1990; Bill and Wicks 1990].

The connector of acceleration sensors can be set in the sensor axis – opposite the base – atop of the housing or at the side of the housing in a radial direction. A

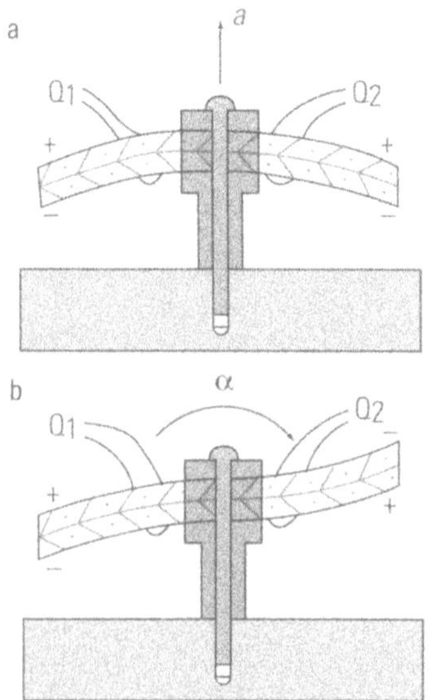

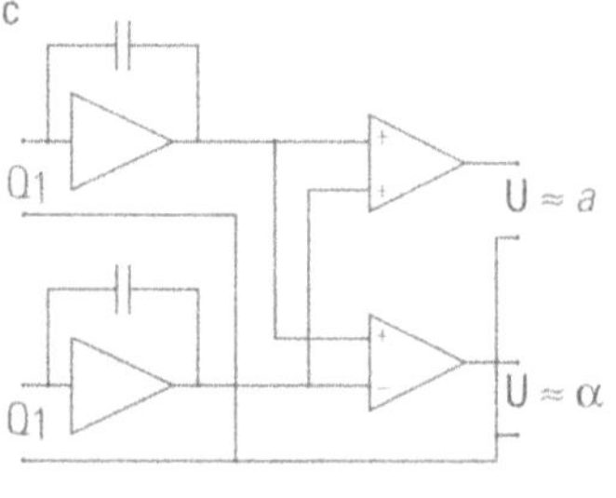

Figure 9.14 Variant of design shown in Fig. 9.13: 2 bending beams are clamped separately onto a mounting post (forming again a double cantilever, each with its own electrodes). Linear acceleration a gives equal outputs (Q_1=Q_2) from both beams (**a**), the sum of these outputs is proportional to a. Angular acceleration α produces equal but opposite outputs (Q_1= Q_2) in the 2 beams (**b**). The difference of these outputs is proportional to α. The built-in electronics (**c**) add and subtract the 2 signals directly and the sensor gives 2 outputs: a and α

lateral connector has the disadvantage that its position after the sensor has been tightened with the specified mounting torque can not be predicted as it depends on the mounting thread. This may cause problems for connecting the cable if space is at a premium at the sensor location. Sensors with an axial connector do not have this drawback, but more space is required in the axial direction.

9.4 Characteristics of Sensors with Piezoelectric Ceramic or Tourmaline Elements

Quartz is the most common transduction element in force, strain and pressure sensors, while for acceleration sensors other materials such as tourmaline and lithium niobate and especially piezoelectric ceramics have found widespread application. The reason is the need for small, light and low-cost sensors, that still have a good sensitivity. Also in measuring vibration, a quasistatic response – impossible to obtain with these materials – is not needed. Sensors with ceramic transduction elements can be extremely miniaturized, as is illustrated by Fig. 9.15.

While sensors with quartz and tourmaline transduction elements have a very good linearity (typically well within ±1%FSO), in sensors with piezoelectric ceramic elements the relationship between measurand and output is not linear, i.e. the linearity of such sensors is only about within ±3 to ±5 %FSO. This corresponds to the specification given by some manufacturers, stating e.g. "Range: ±5000*g*, sensitivity increases about 1 % per 500*g*" which means that the sensitivity is about 10% higher at the upper end of the range than near zero. Therefore, it is important to know at which magnitude of measurand the sensor was calibrated. In critical applications the influence of this non-linear relationship between measurand and output on the accuracy of the measurement must be carefully considered.

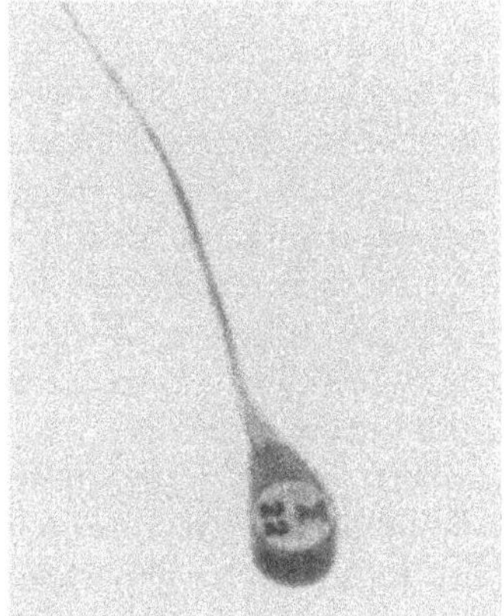

Figure 9.15 Miniature acceleration sensor with a ceramic transduction element for the shear effect. The sensor with a mass of only 0,14g is shown at twice its real size (Courtesy of Endevco)

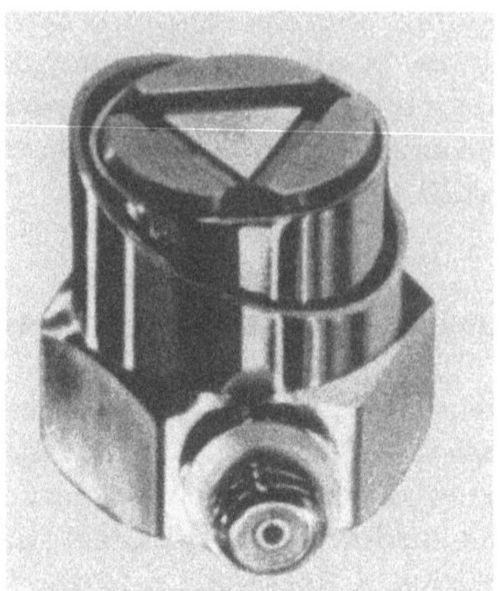

Figure 9.16 Acceleration sensor with 3 radially preloaded seismic masses, each compressing a ceramic transduction element for the shear effect against the triangular center stud. The preload is obtained by a cylindrical sleeve. This design is know as "Delta Shear" (Courtesy of Brüel & Kjær)

Sensors with piezoelectric ceramic elements exhibit hysteresis because all piezoelectric ceramics are constituted of ferroelectric materials which inherently have a hysteresis in the relationship between measurand and output (see chapter 3.7.1). Unfortunately, hysteresis is rarely specified by the manufacturers of ceramic acceleration sensors. Since hysteresis is, in principle, an amplitude-dependent phase shift between output and measurand, it can lead to errors in critical measurements.

Piezoelectric ceramics as well as tourmaline inherently are pyroelectric. This must be considered especially if rapid and substantial temperature changes occur during measurements. The electric charge yielded through the pyroelectric effect appears only on the faces normal to the polar axis. Therefore more and more acceleration sensors are designed with a piezoelectric element that exploits the shear effect, because in ceramics and tourmaline the electric charge from the shear effect appears on faces parallel to the polar axis. In that way, the influence of the pyroelectric effect can be almost eliminated. Examples of such designs are the "Isoshear®" (made by Endevco Corp.) and the "Deltashear®" (made by Brüel & Kjær), as illustrated in Figs. 9.11b and 9.16. Such sensors exhibit similarly low values for the temperature transient error as sensors with quartz elements.

In this context, it must be remembered that certain sensors with quartz elements, especially those with elements cut for the longitudinal effect, show a pseudo pyroelectric effect, although quartz itself is not pyroelectric. This stems from the different thermal expansion coefficients of the materials that are participating in the preloading of the transduction element.

Sensors with ceramic or tourmaline elements have insulation resistance of only about $1 \ldots 100\,\mathrm{G\Omega}$ and are generally not suitable for measuring at very low frequencies, i.e. below about 1...10Hz [Barret 1995]. This is because the time constant in the input circuit of the charge amplifier is also influenced by the insulation resistance of the sensor, as explained in chapter 11.1.2. Moreover, this insulation resistance drops rapidly with increasing temperature and is, at 250°C, about 10MΩ...1GΩ and, at 500°C, only about 100kΩ...1MΩ.

When measuring shocks, a zero shift is sometimes observed [Broch 1970, Brüel & Kjær 1976, Pennington 1965]. Possible causes can be saturation of the amplifier or a nonlinear characteristic of the amplifier, but it may also originate in the sensor itself. The substantial change in the mechanical stress within the preloaded elements – whose elastic properties are different and, in the piezoelectric element, even anisotropic – during the shock can lead to a slight change in the preload of the piezoelectric elements after the shock pulse has ended. This can even happen with sensors that use quartz elements. Another cause for this phenomenon can be the relaxation period of the piezoelectric polarization exhibited by certain piezoelectric ceramics.

9.5 Acceleration Sensors with High Sensitivity

A high sensitivity, i.e. a low threshold, can be obtained by increasing the seismic mass and by using a transduction element with a high piezoelectric sensitivity.

Figure 9.17 shows a design of a sensor with a threshold of below $10^{-5}g$ and a range of $\pm 1g$. The transduction element consists of a stack of piezoelectric ceramic elements, switched in series, and exploiting the longitudinal effect. Combined with the large seismic mass, a sensitivity of 10000pC/g is obtained. The resonant frequency is, due to the large seismic mass, only about 2,5kHz, which is sufficient for applications in geophysics and civil engineering, where the large mass of the sensor of 600g poses no problems either. This sensor has a direct output for connecting a charge amplifier as well as a voltage output from the also built-in impedance converter, which allows to use very long cables.

Basically, the sensitivity could be increased almost without limits by using even larger seismic masses, accompanied – of course – by a decreasing resonant

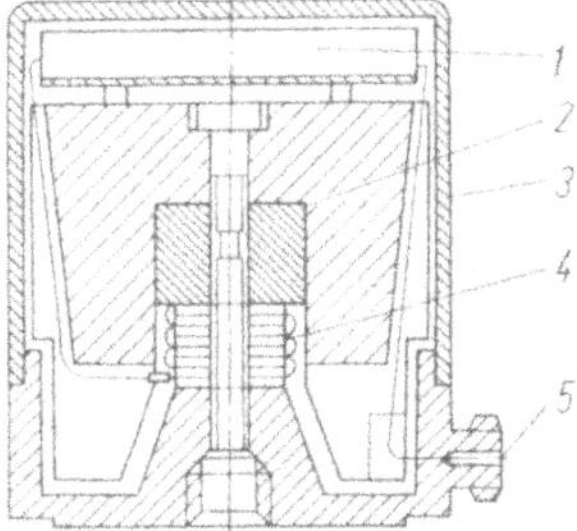

Figure 9.17 High-sensitivity acceleration sensor (Courtesy of Brüel & Kjær) *1* built-in amplifier (is part of the seismic mass!), *2* seismic mass, *3* housing, *4* ceramic transduction element for the longitudinal effect, *5* connector

Figure 9.18 3-component seismometer with a single seismic mass on a 3-component force sensor (Courtesy of Kistler)

frequency. In practice this is rarely done because such sensor would not only become cumbersome to handle, but also difficult to calibrate. Moreover, such sensors would be delicate to handle because of the high risk of overloading them by shocks.

Another interesting example for a high-sensitivity acceleration sensor is the 3-component seismometer shown in Fig. 9.18. A 3-component quartz force sensor (Fig. 6.12) is mounted under preload between the seismic mass and the the base plate of the seismometer. The seismic mass is bell-shaped to bring its center of gravity into the center point of the sensor, essential for minimizing cross talk. This 3-component acceleration sensor with only one seismic mass has the advantage that there is no phase shift between the 3 axes, sometimes encountered in the conventional design combining 3 single-axis sensors for to obtain a 3-component sensor. The range is $\pm 2g$, with a threshold below $0{,}001\,g$, and the frequency range covers 0,1 ... 50 Hz.

9.6 Acceleration Sensors for General Applications

The main field of application for piezoelectric acceleration sensors is measuring oscillations and vibration on machines and structures. The key criteria in selecting an acceleration sensor are range, sensitivity, mass, resonant frequency and operating temperature range.

A large number of sensor types are suitable for standard applications as far as range and operating temperature range is concerned, except for measuring shock above about $20\,000\,g$ and when measuring at temperatures above 250 °C.

For obtaining best results, sensors that offer at the same time a high sensitivity (low threshold), a small mass (minimal mass loading on the measured object) and a high resonant frequency (large frequency range) would be ideal. However, all sensors are a compromise because in acceleration sensors, sensitivity, mass and resonant frequency are closely related.

Figure 9.19 gives an overview on the relationship between sensitivity and mass, and sensitivity and resonant frequency. The graph includes sensors for measuring

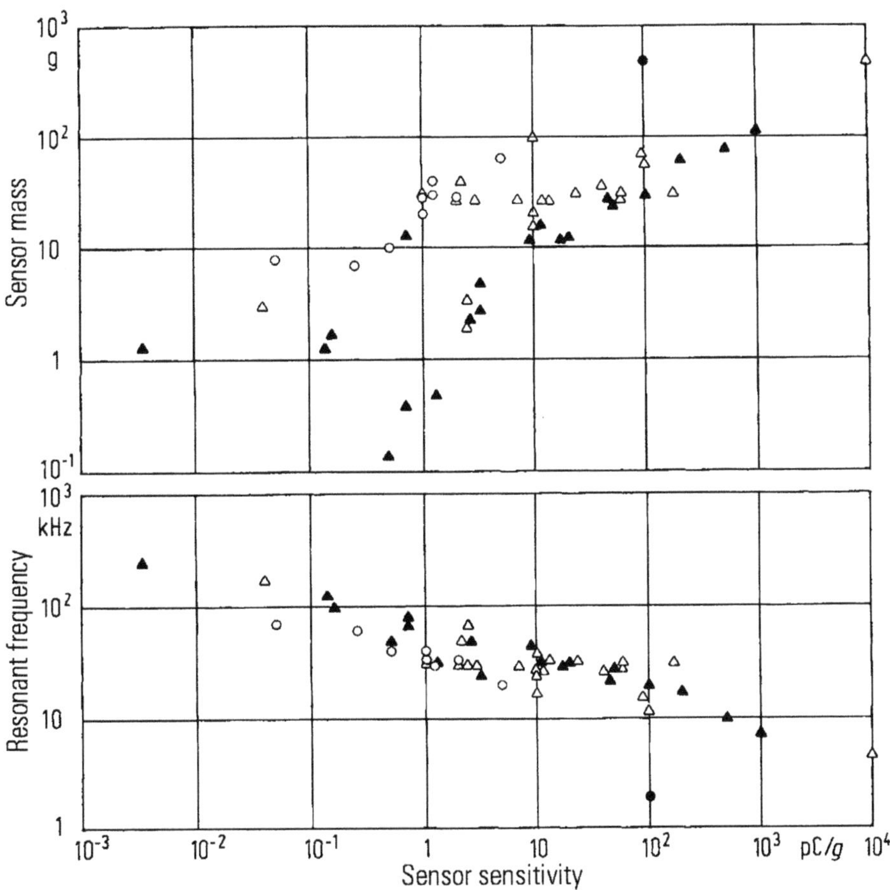

Figure 9.19 Overview of the relationships between sensitivity and mass, and sensitivity and mounted resonant frequencies in acceleration sensors with transduction elements of quartz or piezoelectric ceramic (based on information from Brüel & Kjær, Endevco, Kistler and PCB. ○ quartz element for the longitudinal effect, ● quartz element for the transverse effect, △ piezoelectric ceramic element for the longitudinal effect, ▲ piezoelectric ceramic element for the transverse effect

shock and high-sensitive sensors. The difference in characteristics offered by sensors with quartz elements and those with ceramic elements are clearly visible.

In general, it can be said that

- for a given sensitivity, the mass of a sensor with ceramic elements (especially those for the shear effect) can be up to 100-times smaller than that of a sensor with quartz elements,
- the lightest sensors with quartz elements have a mass of about 0,7 g while, with ceramic elements, masses down to 0,2 g are possible,
- there is little difference in the resonant frequency of sensors with quartz and sensors with ceramic elements, yet – for a given sensitivity – sensors with quartz

elements have always a slightly lower resonant frequency. The highest resonant frequencies achieved with quartz elements are around 125kHz, while sensors with ceramic elements have been made having resonant frequencies of over 250kHz.

Of course, there are characteristics other than the three mentioned above which need to be considered in finding the optimal solution for a given measuring task. These include the range, the operating temperature range, the base strain sensitivity, temperature and temperature gradient error, magnetic sensitivity, acoustic sensitivity (measuring vibration in a very noisy environment), radiation sensitivity, cross talk (also called transverse sensitivity), linearity and hysteresis, maximum transverse acceleration, and so on. Detailed information can be found in [ISA 1995] as well as in the technical literature published by the sensor manufacturers.

Besides acceleration sensors that are connected with a highly insulating cable to a charge or electrometer amplifier, sensors with built-in impedance converter are widely used. This is usually an electrometer circuit (as described in 11.2), especially with ceramic transduction elements, because the inherent high charge sensitivity of piezoelectric ceramic yields – in combination with the small capacitance typical of such elements – a high voltage and hence a high sensor sensitivity. With quartz elements often a miniaturized charge amplifier is built into the sensor because the lower charge output of quartz would not give a sufficient voltage across the capacitance of the transduction element. Such sensors offer the practical advantage that – instead of a highly insulating and low-noise cable – an ordinary cable, even without shielding, can be used as the output has a very low impedance. Moreover these circuits are uncritical in an industrial environment, i.e. they are virtually insensitive to electric and magnetic fields or other electric interference.

Sensors with built-in electronics have certain limitations and drawbacks. The range of the sensor is fixed and can not be changed later. Also the lower frequency limit is fixed once and for all. A very low frequency limit built into the sensor has the disadvantage that – after e.g. a temperature transient – it takes a certain time (corresponding to the necessary long time constant) before the output of the sensor has returned to zero. A shorter time constant reduces this effect but leads to a higher lower frequency limit. Also the operating temperature range is limited by the built-in electronics to about 40…125°C (with special electronic components up to about 165°C).

9.7 Acceleration Sensors for Measuring Shock

Vibration has the nature of a periodic signal, while shock is clearly an aperiodic phenomenon, characterized by a rapidly and steeply rising front and an equally rapid return to zero. In an idealized way, a shock can be considered as a half of a sinusoidal oscillation. The duration of shock is usually in a range of a few μs to several hundred ms, which requires sensors with a very high natural frequency for rendering correctly the very short rise times without distortion, overshoot or ringing. At the same time, the lower frequency limit must be low enough for avoiding undershoot at the end of the shock. Especially when the signal is to be integrated, a near DC response is needed, i.e. an external charge amplifier set to DC-mode must be used. For precisely measuring shock, a sufficiently wide frequency response is a sine qua non – a requirement often overlooked in practice.

In vibration the peak-to-peak amplitudes are rarely more than about 1000 *g*. However for measuring shock, a range of over 10000 *g* is needed, in some cases up to 100000 *g*.

In measuring shock, the sensors should always be mounted in such a way, that the shock wave produces compression in the contact interface between sensor and object. Otherwise, the mounting stud or bolt would be loaded by tension, resulting in a softer coupling and a reduced rise time of the system. Also the internal preloading element of the sensor – many shock sensors are of the compression design with piezoelectric elements exploiting the longitudinal effect – would be stressed by tension, with the same effect on the measuring result. For this reason, high-range shock sensors often have an asymmetric measuring range, e.g. –20 000 ... 100 000 *g*.

Assuming a charge amplifier with a sufficiently wide frequency response, it is the natural frequency and the ringing frequency of the sensor that determine the error in reproducing a shock pulse, while in measuring vibration it is the resonant frequency that governs the upper frequency limit of the system.

Shock sensors should be free of zero shift after having been loaded by a shock. As already explained in 9.4, this problem exists especially in sensors using a piezoelectric ceramic element. Although quartz itself is free of such effects, zero shifts in sensors with quartz elements are often observed, too. The origin of such zero shifts after a high transient loading is a slight change of the stress pattern in the interfaces between quartz elements with anisotropic moduli of elasticity and the usually not anisotropic sensor material (preloading elements, electrodes, etc.) within the path of mechanical preload. Such change in stress pattern influences through the transverse piezoelectric effect in the quartz elements the output signal, observed as a zero shift. This effect can be minimized by sensor parts that are machined to very close tolerance and by appropriate design of the sensors.

More recent designs use quartz elements exploiting the shear effect and have a built-in miniaturized charge amplifier. These sensors are free of zero shift, have a very high resonant frequency (>100 kHz), have a lower frequency limit of 0,5 Hz and offer a symmetrical range of up to ±100000 *g* with a sensitivity of 0,05 mV/*g*, yet have a mass of only 4,5 g (Fig. 9.20). They exhibit the excellent linearity inherent of quartz, i.e. better than ±1 %FSO. A still higher resonant frequency of 250 kHz, combined with a very low mass of only 1,3 g, is featured by the sensor shown in Fig. 9.21. It uses ceramic shear elements with a linearity of only within about ±5 %FSO (the sensitivity at the upper end of the range at 100000 *g* is about 10% higher that in the partial range at 10000 *g*). Its lower frequency limit is rather high, i.e. about 5 Hz, and the sensitivity is quite low at 0,0035 pC/*g*, which corresponds to 0,0035 mV/*g* when connected to a good charge amplifier.

For precise measuring of shock the type of sensor to be used must be carefully selected. When the amplitude and also the integral are needed, sensors not only

Figure 9.20 Acceleration sensor for measuring shock up to ±100 000 g with a quartz element for the longitudinal effect (Courtesy of Kistler)

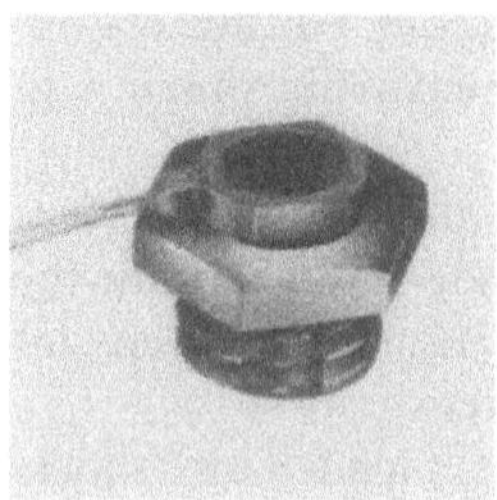

Figure 9.21 Acceleration sensor for measuring shock up to ±100 000 g with piezoelectric ceramic elements for the shear effect and built-in amplifier (Courtesy of Endevco)

with a high linearity but especially with a very low lower frequency limit are required. Quartz sensors offer the best solution, especially sensors with charge output connected to a charge amplifier with a good near DC response, i.e. a quasistatic response [Broch 1970, Pennington 1965].

9.8 Acceleration Sensors for High Temperature

For measuring acceleration at temperatures above about 250 °C, only sensors with tourmaline, lithium niobate and special piezoelectric ceramics (e.g. lead meta niobate) can be used. Organic materials such as PTFE will not work above 250°C and ceramics, glass or metal oxide must be used as insulation material. In measuring vibration above frequencies of about 10 Hz, the relatively high lower frequency limit caused by the decrease of the insulation resistance with increasing temperature and the pyroelectric effect in some of these materials rarely pose problems in practice.

The piezoelectric sensitivity of the transduction element drops with increasing temperature and it is necessary to calibrate at elevated temperature, too. The special calibration equipment required is usually designed by the respective sensor manufacturers.

Figure 9.22 High-temperature acceleration sensor with a tourmaline element for the longitudinal effect for monitoring vibrations e.g. in jet engines (Courtesy of Vibrometer)

Figure 9.23 High-temperature acceleration sensor with a lithium niobate element for the shear effect. The welded-on metal cable has an insulation of magnesium oxide (Courtesy of Endevco)

A high-temperature vibration sensor with a tourmaline transduction element for monitoring engine vibration in aircraft is shown in Fig. 9.22. Designed for continuous operation (MTBF is more than 100000h) up to 780°C, this type of sensor has been used successfully for over 30 years by nearly all airlines around the globe and is an excellent example for the high stability and reliability inherent of piezoelectric sensors. Another type of sensor uses a lithium niobate element exploiting the shear effect and also works up to about 760°C (Fig. 9.23). Applications for high-temperature vibration sensors also include monitoring tasks e.g. in nuclear power stations.

The insulation resistance of tourmaline and lithium niobate is only about 50kΩ at temperatures above 600°C and therefore charge amplifiers with specially modified input stages must be used (11.5.8)). The lower frequency limit is then at around 20 to 50Hz only.

Interestingly enough, no acceleration sensors with quartz elements have been made so far, although this would be quite feasible. Contrary to what is sometimes stated (e.g. [Grave 1962, p 57] and [Neubert 1975, p 259], quartz does not loose its piezoelectric properties above 573°C, the temperature of the phase change from the α-phase into the β-phase. As explained in 4.2.8, β-quartz is piezoelectric, too, and could be used within the temperature range of about 600 to 800°C.

Figure 9.24 Acceleration sensor with quartz shear elements sensor for low-temperature applications (Courtesy of Kistler)

Other possibilities are offered by more recently introduced materials such as crystals of the CGG group (see 3.5) and gallium orthophosphate (see 3.4).

9.9 Acceleration Sensors for Low Temperature

Applications in space and ESS (environmental stress screening) often include measuring at very low temperatures down to below 200 °C. It may be noted here that the value of 196°C stated by most manufacturers as the lower limit of the operating temperature range of a sensors simply corresponds to the boiling point of liquid nitrogen, which is a readily available medium for verifying the proper operation of a sensor at low temperature.

Quartz is the preferred transduction element for low-temperature sensors because the piezoelectric sensitivity of quartz drops only little at these temperatures. A miniature sensor with built-in electronics and a mass of only 1,6 g is illustrated in Fig. 9.24. Its sensitivity is 10 mV/g and it features the high linearity of quartz elements, exploiting here the shear effect.

9.10 Acceleration Sensors for Modal Analysis

Modal analysis – the study of the dynamic behavior of structures – has become an indispensable tool for optimizing and verifying the design of structures in vehicles, machinery, bridges and many other fields [Ewins 2000]. Often, a large number of vibration sensors is needed for a complete analysis and, therefore, size, mass – and price are obviously critical. The required specifications are not extreme: a range of about $\pm 2g$ to $\pm 50g$, a frequency response from below 1 Hz up to a few kHz, a sensitivity of around 500 to 1000 mV/g and a mass of a few g is sufficient for most applications.

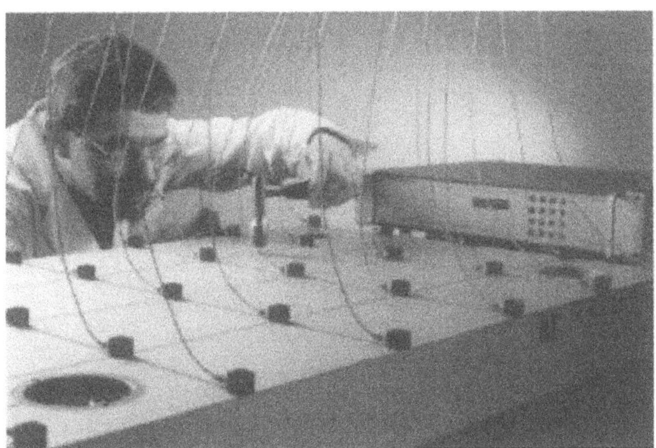

Figure 9.25 Modal analysis of a top plate of a force plate for biomechancis (see Fig. 6.29). The plate is excited with an impulse hammer (see Fig. 9.27). Sensors arranged in a geometric pattern pick up the response of the plate to the measured impulse given with the hammer (Courtesy of Kistler)

Most sensors designed for modal analysis use piezoelectric ceramic elements, combined with a built-in charge amplifier. By using a charge amplifier, the high piezoelectric sensitivity inherent of ceramics can be fully exploited. The amplifier circuitry is designed to be excited by a constant current (usually a few mA) and the sensor presents to the current source a resistance which varies proportionally to the measurand. A precise resistor switched in series into the 2-wire constant current circuit allows to pick the voltage drop across it as a voltage output, again proportional to the measurand. Some sensor use an impedance converter (really an electrometer amplifier) instead of a charge amplifier in a similar arrangement. This 2-wire technique – known as Piezotron®, ICP® and other trade marks – is described in 11.3.

Modal analysis generally requires a large number of sensors (see Figs. 9.25 and 9.26). Therefor the cost of the sensors and amplifiers is critical, which is another reason why sensors with built-in electronics are the preferred choice.

For exciting structures to perform modal analysis, instrumented hammers (so-called impulse hammers) are used (see Fig. 9.27) [Bill 1998]. The head of the hammer is fitted with a quartz force sensor, on which impact tips can be attached. Depending on the nature of the structure, a tip of appropriate hardness (steel, plastic, soft PVC, rubber, etc.) is chosen. The mass of the hammer head can be just 100 g for exciting small structures while for heavy structures, heads with masses up to several kg are used. In some applications, so-called *mechanical impedance heads* (Fig. 9.28) are used, providing phase-matched force and acceleration signals at the driving point of excitation of the structure. Another way of exciting small structures is e.g. a hand-held exciter (Fig. 9.29) that can be fitted with a force sensor and be driven with an input signal of any shape (sinus, random, square wave, saw tooth, etc.).

Figure 9.26 Modal analysis of a car body. Again, a large number of sensors is needed to obtain a complete response pattern of the structure to a dynamic excitation (Courtesy of Kistler)

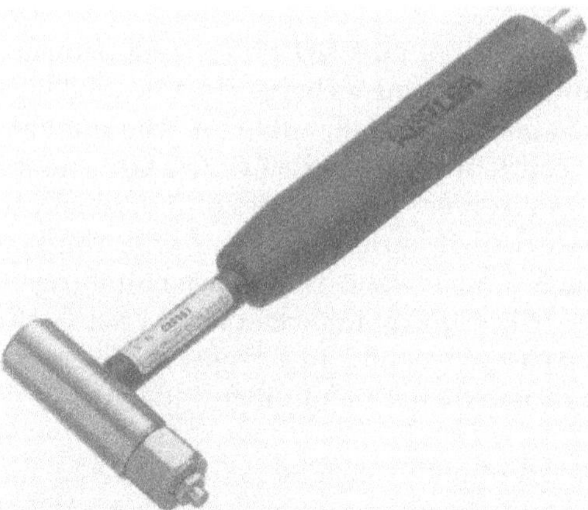

Figure 9.27 Impulse hammer with quartz force sensor and steel impact tip. The sensor output is picked up at the connector at the rear end of the handle, so the cable will not be damaged accidentally when working with the hammer (Courtesy of Kistler)

Figure 9.28 Impedance head (force and acceleration sensors combined in one housing) for measuring driving point input data in modal analysis (Courtesy of Kistler)

Figure 9.29 Hand-held exciter for exciting and testing small and medium size objects. Freqeuncy range is about 45 Hz to 15 kHz (Courtesy of Brüel & Kjær)

Figure 9.30 Triaxial miniature quartz sensor with wide frequency response (resonant frequency: 80 kHz) (Courtesy of Kistler)

9.11 Triaxial Acceleration Sensors

In many applications, sensors that allow to determine the direction of the acting acceleration are needed. This can be done with triaxial sensors. There are two solutions: mount 3 uniaxial sensors on a mounting block, one sensor for each direction, or use a triaxial sensor.

Using 3 sensors on a mounting block has the advantage that practically any type of sensor can be used. However the mass of 3 sensors and of the mounting block is quite high, which excludes this solution when mass loading is critical, i.e. when the measured object is a light structure.

Triaxial sensors combining 3 sensors in one housing are available in a number of designs and can achieve very low masses (e.g. the model 23 of Endevco has a mass of only 0,85 g). Some special designs of triaxial sensors use only one seismic mass for all 3 directions, such as the ThetaShear® design of Brüel & Kjær. Figure 9.30 shows a miniature triaxial quartz sensor with a mass of only 2,5 g.

When mass is not critical, triaxial acceleration sensors can also be designed around 3-component force sensors (Figs. 9.18 and 6.12).

9.12 Calibratiion of Acceleration Sensors

Calibrating piezoelectric acceleration sensors is a demanding task and requires specialized equipment. The basic problem is to expose the sensor to a precisely known acceleration. A steady acceleration – which can e.g. be produced with a centrifuge – is not very suitable for piezoelectric sensors because of the limited static response obtained from quartz sensors. Sensors with piezoelectric ceramic elements can definitely not be calibrated in a quasistatic way. Therefore one has to resort to periodically (usually sinusoidally) varying acceleration or a shock-like excitation.

The most widely used method employs electrodynamically driven oscillating tables – commonly called shakers – with which a sinusoidally varying acceleration can be produced. While it is very easy to determine the frequency of the oscillation, it is very difficult to measure the amplitude of the oscillation precisely. This method – also known as the absolute method – requires that the amplitude be measured directly. Formerly, measuring microscopes were used to read the peak displacement of a mark on the vibrating sensor or table. More accurate results can be obtained with a laser interferometer. Although the measuring error can be kept within ±0,5%, such sophisticated equipment is not readily available and therefore the most commonly used way of calibrating is by the comparison method. Here, a reference sensor – which may have been calibrated by the absolute method – is used as reference with which the sensor to be calibrated is compared.

The preferred type for calibrating by comparison is an acceleration sensor of the "hanging seismic mass " design (Figs. 9.9d and 9.31). This reference sensor is attached to the table of a shaker and the sensor to be calibrated screwed on top of it, which assures the best coupling between the two sensors, i.e. both sensors will move in exactly the same manner. This method is know as "back-to-back" calibration and allows to calibrate within an error of less than about ±2%, under very good condition and – if done very skillfully – even within ±1%. If the two output signals are added with opposite phases, the gain of the amplifier of the

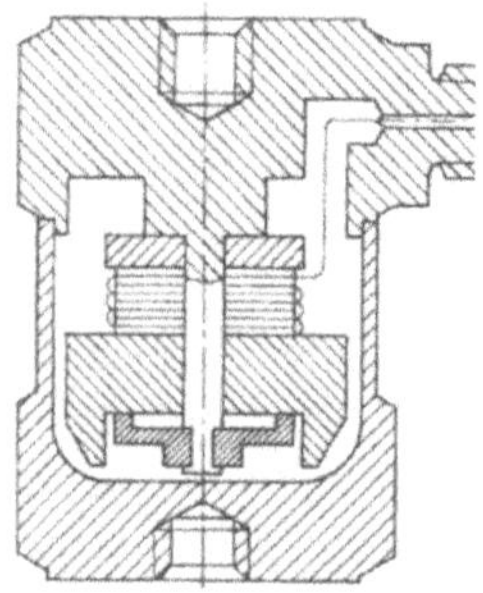

Figure 9.31 Working standard acceleration sensor with quartz elements for the longitudinal effect for calibrating acceleration sensors by the back-to-back method (Courtesy of Brüel & Kjær)

sensor to be calibrated can be adjusted until the differential signal is at a minimum, in the ideal case zero, and the sensitivity determined.

Electrodynamic shakers can be used in a frequency range of about 30 Hz to 20 kHz (some special designs go up to around 50 kHz) and are able to produce amplitudes up to ±50 to ±100 *g*. For lower frequencies, hydraulic shakers can be used. An essential quality characteristic of a shaker is the ability to perform a truly rectilinear motion normal to the mounting surface of its table. Also the table should move exactly parallel, i.e. there must be no tilting during the motion. Usually acceleration sensors are calibrated at certain fixed frequencies, preferably not multiples of the mains frequency in order to minimize electric noise.

Frequency response and resonant frequency can also be determined with a shaker. A reference sensor attached to the table serves to control the driving circuit of the shaker so that it keeps the vibration amplitude constant, irrespective of the frequency. Sweeping through the frequency range and recording the output amplitude as a function of the frequency yields the frequency response curve. If the range of the shaker is sufficiently wide, it is even possible to detect the resonant frequency, i.e. the measurand frequency at which the sensor responds with maximum output (see definition in 4.2.2). This frequency is also called "mounted resonant frequency", provided the mounting of the sensor is identical with the mounting when used (mounting surface quality, material of mounting screw or stud, mounting torque, etc.). Fig. 9.8a shows a typical recording of the frequency response of a sensor, recorded at a constant amplitude of ±1 *g*.

Another method for calibrating is the drop tube. At the bottom of the vertically mounted tube is a quartz force sensor. The accelerometer to be calibrated is screwed onto a cylindrical weight and the total mass is determined by weighing. The weight with the attached sensor is then dropped on the force sensor, guided by the tube, and the output signals recorded. The acceleration a can be calculated from the measured impact force F through the relation $F/m = a$. Peak amplitudes of about 5000 *g* can reached and therefore this method is particularly suitable for calibrating shock sensors. A damping pad must be inserted to prevent the mass to drop metal-on-metal on the force sensor. By choosing a pad of suitable shore hardness, clean half-sine pulses with a minimum of high-frequency noise can be obtained. When using a very hard pad, or even no pad at all, the signal will be noise covering a very wide frequency range and this usually excites the natural frequency of the sensor, too.

The natural frequency of a sensor can often be found also by the exploiting the converse piezoelectric effect. An AC voltage is applied to the transduction element via the connector of the sensor and the frequency varied over a sufficiently wide range (usually about 10 to 200 kHz) while keeping the voltage amplitude constant. When going through the natural frequency of the sensor, the electric impedance of the transduction element will show an abrupt change that can be detected by appropriate instrumentation [Pennington 1965].

The natural frequency, especially of acceleration sensors, is usually higher than the resonant frequency of the sensor determined on a shaker. The shock method excites the natural frequency of the transduction element, i.e. the free, not forced oscillation of the transduction element (4.3.2). The resonant frequency found with the sensor mounted on a shaker is determined not only by the spring (transduction element)/mass (seismic mass) system but also by the elastic properties of the sensor housing, especially the coupling to the mounting surface. Therefore the resonant frequency of acceleration sensors is often called "mounted resonance frequency", but the type of the mounting must be specified (4.3.4).

Acceleration sensors are a good example of sensors in which natural and resonant frequency can be distinguished very clearly (see the precise definitions in 4.3.2). Depending on the type of sensor design, the natural frequency of an acceleration sensor can be higher by a factor of about 1,5 to 2 than the (mounted) resonant frequency, depending on the sensor design. For practical work, the always lower resonant frequency is usually determining the upper usable frequency limit.

Another parameter is the cross talk (also called transverse or cross sensitivity). For measuring it, the sensor is attached to a cylinder that can be rotated about its axis and is held by a metal block with the axis normal to the direction of the shaker motion. The complete testing fixture (block with cylinder and sensor) should be mounted in such a way that its center of mass is in the centerline of the shaker. Otherwise, the eccentric center of mass would induce tilting motions to the shaker table, falsifying the cross talk measurement. A perfect sensor would give no output when vibrated normal to its axis, no matter in which direction. Well designed precision sensors have a cross talk of less than ±1% while general purpose sensors often have dross talk of ±2% or even up to ±5% and more, a point to remember when selecting a sensor for precision measurements. Cross talk normally is not the same in all directions, but shows a maximum in one direction and a minimum in a direction normal to the first one. Some manufacturers mark the direction of minimum cross talk on the sensor housing so it can be oriented appropriately in critical applications.

Strain sensitivity (4.2.4) can be determined by mounting the sensor on a flexing beam, fitted with strain gages that indicated the actual strain in the mounting area of the sensor. The strain sensitivity (often called base strain sensitivity, too) heavily

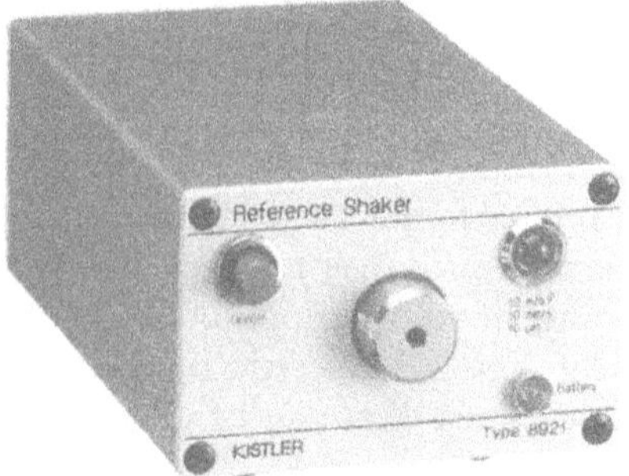

Figure 9.32 Reference shaker for field checking the calibration of acceleration sensors with a mass up to 300gat a fixed frequency of 1000rad/s (159,2Hz) and a constant amplitude of $10\,m/s^2$ (±3%) (Courtesy of Kistler)

depends on the type of sensor design and must be considered especially when measuring on thin and light structures.

More details on the methods to determine the various characteristics of acceleration sensors can be found in [Brüel & Kjær 1976; ISA 1995; Pennington 1965] and in publications by the manufacturers.

Calibration shakers are usually not available in the field. Portable reference shakers (Fig. 9.32) can be used to verify the calibration. Such shakers produce vibration at a fixed frequency (usually 159,2 Hz, corresponding to 1000 rad/s) and at an electronically regulated fixed amplitude, independent of the sensors mass (up to the specified limit) and allow to check whether the output of the sensor is still within e.g. ±3 % of its original value.

10 Acoustic Emission Sensors

10.1 Quantity and Units of Measurement

The quantity "acoustic emission" can only be defined in a qualitative way. An excellent formal definition can be found in [Miller and McIntire 1987]: "Acoustic emission is the class of phenomena where transient elastic waves are generated by the rapid release of energy from localized sources within a material, or the transient waves so generated". Acoustic emission (often just called "AE") can be looked at as a vibration with an extremely small amplitude (in the order of nm) and of very high frequency (in the order of 10 kHz to over 1 MHz) of particles in a solid material. It can also be described as a sound wave originating in and traveling through a solid (the term "structure-born sound" to describe AE is also employed by some authors). The units usually used for measuring vibration or acceleration (see 9.1) are sometimes used in AE work, when acceleration sensors with a high sensitivity are used to capture AE in certain applications. Units used in measuring (acoustic or sound) pressure (see 8.1) are also suitable to quantify AE, especially if the sensor – a pressure sensor of high sensitivity, such as a hydrophone – is coupled by a fluid to the test object or used in underwater investigations.

AE causes a dynamic motion of the test object's surface. Either this motion can be captured as a function of time (the AE sensor is then acting as a displacement sensor) or the velocity of the surface motion is measured (then the AE sensor acts as a velocity sensor).

Within the scope of this book the measurand "acoustic emission" is only considered as a parameter used in process control and monitoring applications. For the vast field of nondestructive materials testing, the reader is referred to [Miller and McIntire 1987] and the references cited there.

Depending on the method of calibration and the type of sensor, the sensitivity of an AE sensor is given in V/µm when the sensor is used to measure surface displacement or motion, or in V/(mm/s) when the sensor is intended to measure surface velocity. Response to continuous excitation may also be characterized in terms of pressure, such as V/µbar. Sensitivity is usually plotted against frequency and the amplitude is expressed in $dB_{ref\ 1\,V/m}$ or $dB_{ref\ 1\,V/(m/s)}$.

10.2 Background of AE as Process Parameter

While AE is a well established parameter in nondestructive testing, it is still little known in general engineering and process control. Therefore, a brief historic review may familiarize the reader with the particularities of AE and stimulate ideas for new applications in process control.

AE has been a naturally occurring phenomenon since primeval time. Even before man appeared on earth, animals living in trees must have learned to deduct from the cracking sound emitted by a breaking branch that failure was imminent. Similarly primitive man was warned by the creaking sound from a trunk serving as a bridge or from a beam supporting the roof that its load capacity was about to be exceeded.

Timber emits cracking sounds when loaded – witness the popular saying: "The beams are cracking!" – and coal miners relied from the beginning, when they used timber to stabilize the tunnels and pits in mines, on this type of AE as a reliable warning that structural failure followed by a collapse or cave-in was imminent. The old miner's rule "silver fir cracks earlier than spruce fir" demonstrates that differences in the AE sensitivity of various types of wood had been recognized early. For that reason, pit props were preferably made of silver fir and probably were the first AE sensors that served as a warning system to save human lives.

Apparently the earliest attempt to measure AE from wood was made by F. Kishinouye (see also 2.5) and published in 1934 in Japanese [Kishinouye 1934]. An English translation became available only in 1990 [Kishinouye 1990]. Kishinouye also tried to measure AE from mechanically loaded rock, however did not succeed due to the lack of suitable equipment. These studies were made with the intention to detect earthquakes, i.e. to use AE as a warning indicator. More details on the idea of using AE from wood as a safety monitoring parameter or an early warning system can be found in [Drouillard 1990 and 1994] with a wealth of further references.

Probably the earliest use of AE for process control in the largest sense of the term was by makers of hard-fired pottery. The use of fired clay for utilitarian, domestic purposes dates back over 10000 years to the Neolithic Period. The oldest piece of fired pottery has been dated at about 7900 BC and was found in Henan Province, China. From the beginning, potters most certainly must have observed that fired clay vessels, when cooling off too fast in the kiln, always emit cracking sounds, and that such pieces invariably turn out to be defective, i.e. sooner or later, they are likely to fail structurally when put to their intended use.

AE from metals most likely was first observed when working pure tin, which became possible about 3700 BC in Minor Asia, where man seems to have learned to smelt pure tin for the first time (begin of copper making) [Aitchison 1960]. Pure tin gives off a harsh screeching or crackling sound when subjected to plastic deformation, resulting from the mechanical twinning in the metal. This type of AE – which became known as "tin cry" – appears to have been documented for the first time in the 8th century by the Arab alchemist Jabir ibn Hayyan, Abu Musa (also known as Geber) in his book "*The Sum of Perfection or the Perfect Magistery*" which was published in Berne (Switzerland) in a Latin translation in 1545 and in an English translation in 1678 [Geber 1928]. In the early 20th century, AE from an increasing variety of metals, crystals and nonmetallic materials were described. It is now well established that nearly all kinds of metal working are associated with AE which, therefore, has the potential of becoming another meaningful process control parameter.

In 1948, Warren P. Mason succeeded for the first time – exploiting the piezoelectric effect in quartz – in measuring shifts of 0,1 nm in the surface of metals, occurring in a few µs only [Mason et al 1948]. Utilizing the piezoelectric effect in Rochelle salt, D. J. Millard measured the AE during twinning of monocrystalline cadmium under mechanical load in 1950 [Millard 1950]. In the same year, J. Kaiser described the relationship between the strain-tensile stress curve of different materials and the intensity of the associated AE, i.e. the various frequencies noted for the various stresses to which the specimens were subjected [Kaiser 1950]. This was to become the foundation of modern AE measuring techniques and the new technology of nondestructive materials testing emerging in the sixties. Further development confirmed that the ideal measuring principle for measuring AE was the piezoelectric effect, indeed.

The most significant discovery Kaiser made was the irreversibility of the phenomenon "acoustic emission", which is now known as the *Kaiser effect*. AE, sometimes also called "structure-born sound", inherently is an irreversible phenomenon, i.e. it indicates an irreversible change in the structure of the material. This is exactly in the nature of most industrial processes and therefore, AE will certainly find an increasing number of applications in the field of process monitoring and control, too.

10.3 Design of AE Sensors

AE sensors are almost always of the piezoelectric type. Sensors designed specifically for measuring AE are similar to acceleration sensors except that they do not need a seismic mass, but rather a special backing or damping material applied directly to the transduction element (Fig. 10.1). The transduction element is usually a piezoelectric ceramic disk or cylinder with a thickness of a few mm. Similarly the diameter can be as small as 1 mm, but usually is around 5 mm. The base of the sensor is coupled to the test object by a thin layer of a couplant or an adhesive bond. Some sensors can be pressed onto the surface of the test object with a mounting bolt or with a magnetic clamp.

The backing material is often made of epoxy mixed with high-density (e.g. tungsten) particles before curing. The aim is to let the acoustic waves pass readily into this backing material with as little reflection as possible, and to absorb the wave's energy by scattering the wave on the embedded particles. The backing pad serves to load the transduction element to make it less resonant and to obtain a wider frequency response of the sensor (wideband-type sensors). For special

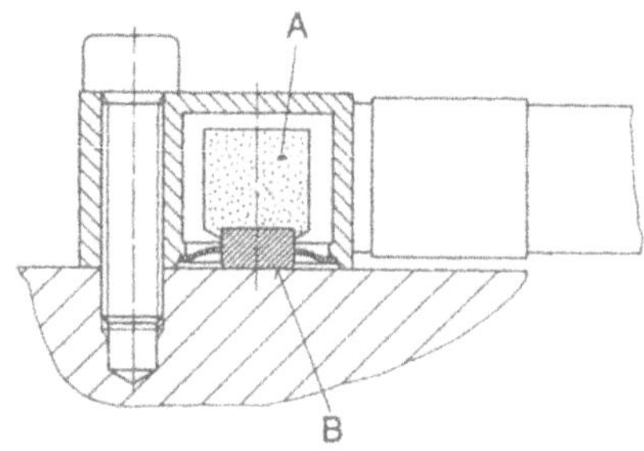

Figure 10.1 Design of an AE sensor (Courtesy of Kistler)

applications, where the frequency band of the phenomenon to be studied is very narrow and at a known value, sensors are designed with a resonant frequency matching that frequency and providing a much higher output signal (resonant-type sensors).

Some AE sensors have a direct output to be connected to an external charge amplifier or electrometer amplifier, but a design with a built-in amplifier is often preferred because the influence of the cable – especially if it is a long one – on high frequency response, amplitude and noise is eliminated.

The topic of AE sensor design – especially for use in nondestructive testing – is exhaustively treated in [Miller and McIntire 1987, sections 5 and 15]. For in-process monitoring, AE sensors must satisfy a number of criteria, i.e. they should

- have a wide and flat frequency response,
- have a high AE sensitivity,
- provide good, reproducible and stable coupling to the mounting face,
- be insensitive to low-frequency noise (inherent high-pass characteristic),
- be insensitive to electric and magnetic fields (ground isolation, low-impedance output),
- have a low mass (minimal mass loading effect),
- be as small as possible,
- be very rugged (suitable for duty in harsh industrial environments, i.e. at least protection class IP65 or IP67), and
- allow easy mounting, especially for retrofitting existing machinery).

Fig. 10.2 shows an AE sensor that meets these requirements in an optimal way and is designed for applications in an industrial environment.

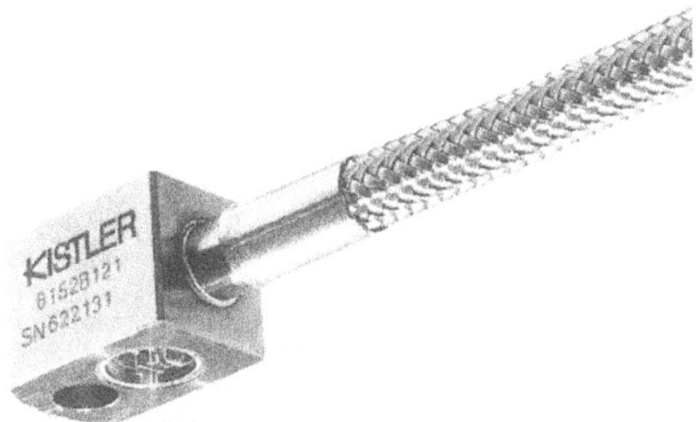

Figure 10.2 AE sensor for industrial use (Courtesy of Kistler)

10.4 AE Sensors for Process Monitoring

In a growing number of processes – besides the established measurands force, strain, pressure and vibration – AE is being discovered as a new parameter potentially useful for monitoring and possibly controlling. Examples are:

- metal cutting (tool wear and breakage)
- metal forming such as stamping and deep drawing
- extruding plastic melts, especially filled melts
- indicating the stress level in bolts
- montoring of welding processes
- machinery condition monitoring and incipient failure detection (e.g. cavitation in pumps, monitoring of bearings, failure of seals, etc.)
- monitoring of aircraft structures

10.5 AE Sensors Applied to Metalworking

AE sensors can be used to detect a variety of phenomena and parameters in metalworking, as illustrated by two examples.

When stamping sheet aluminum (e.g. embossing a letter), the stamping force can be recorded. However there is virtually no difference between untreated and anodized aluminum. If AE is recorded too, there is a striking difference: only the anodized sheet produces a clearly measurable AE signal (Fig. 10.3). An obvious application is quality control in minting coins where measuring force alone does not allow to detect flaws.

In metal cutting, especially with fully automated machine tools, immediate detection of breakouts or complete breakage of tools is of prime importance. Although such tool failures can often be detected as changes in the cutting force by single or multicomponent force sensors, the example shown in Fig. 10.4 illustrates

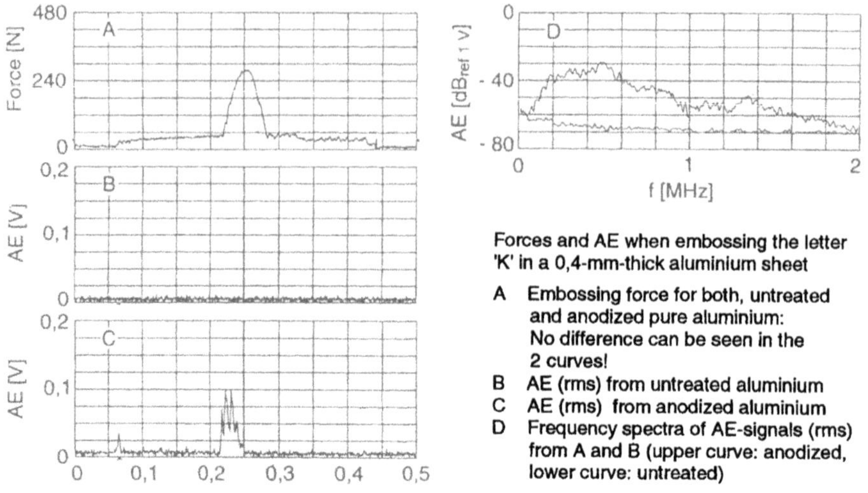

Figure 10.3 Forces and AE when embossing the letter "K" in a 0,4 mm-thick sheet of aluminum (Courtesy of Kistler) **A** Embossing force for both, untreated and anodized pure aluminum. No difference can be seen between the two materials. **B** AE (rms) from untreated aluminum; **C** AE (rms) from anodized aluminum; **D** Frequency spectra of the AE signals (rms) from **B** and **C** (upper curve: anodized, lower curve: untreated)

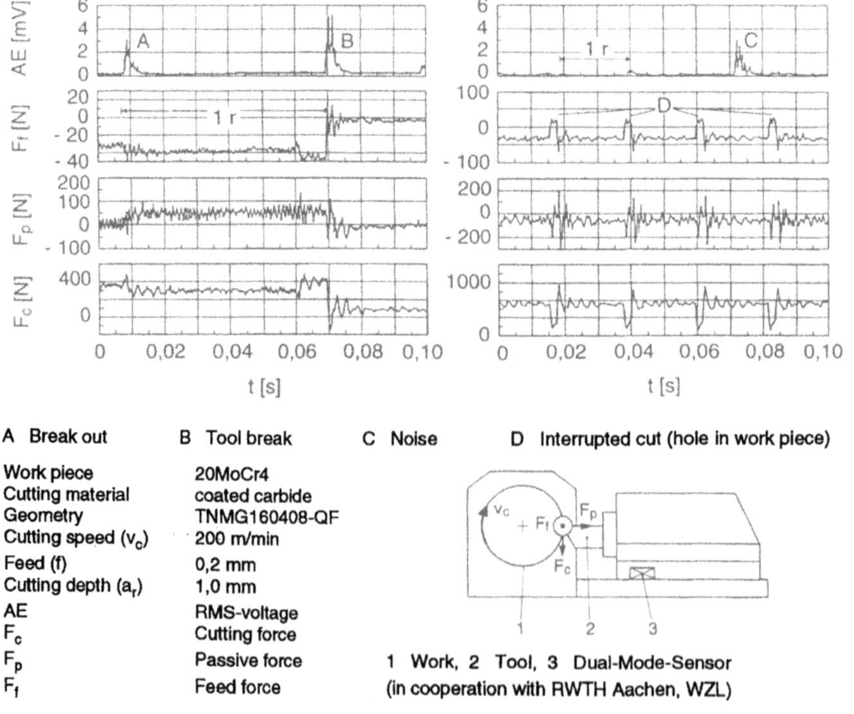

Figure 10.4 Cutting force and AE in turning (Courtesy of Kistler)

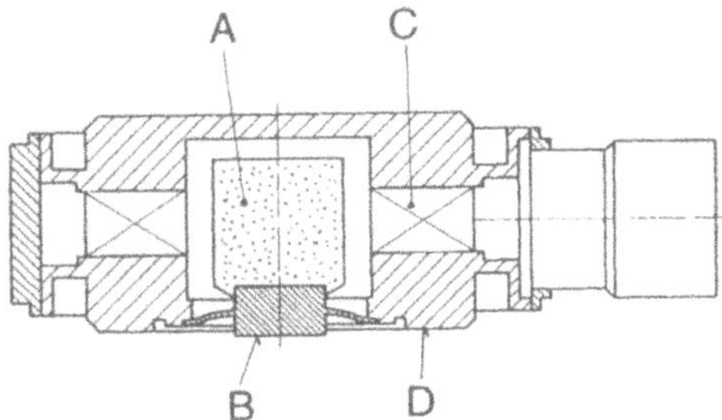

Figure 10.5 Dual-mode sensor for AE and 3-component force (cross section) (Courtesy of Kistler) **A** AE sensing element, **B** AE coupling face, **C** 3-component force sensor, **D** force coupling face

how much more clearly the AE signal does indicate such events. Moreover interrupted cutting (e.g. across a hole or slit in the work) will naturally show up in the cutting force. However, there is no AE signal generated as long as the cutting edge of the tool remains intact. Therefore, monitoring AE is more useful in this situation as only tool breakage needs to be detected, regardless of whether the cut is continuous or interrupted [Marschall and Gautschi 1991].

For these experiments, a novel dual-mode sensor, combining an AE sensor and a 3-component force sensor in one housing was used (Fig. 10.5).

The critical point in using AE sensors is to find the best location for the sensor. This can usually only be found through field tests, searching empirically for the clearest and most relevant AE signal.

10.6 AE Sensors for Nondestructive Testing

The technique of nondestructive testing is a vast specialized field of its own and it would be far beyond the scope of this book to try to cover it, too. Therefore the reader is referred to [Miller and McIntire 1987] where this topic is treated exhaustively and a wealth of references can be found.

10.7 Calibration of AE Sensors

Calibrating sensors for AE is not as straightforward as for other measurands. AE is a measurand of a very complex nature, occurring in a large variety of materials. A meaningful calibration seems to impose that the sensor is calibrated on the same type of material as it is intended to measure on later.

There are essentially two ways of determining the sensitivity of an AE sensor:

- Primary calibration: the surface displacement is measured directly (e.g. by laser interferometry) as described in [ASTM 1986; Liebig 1998]
- Secondary calibration: the sensor to be calibrated and a reference sensor (serving as working standard) are mounted in a symmetric arrangement on the

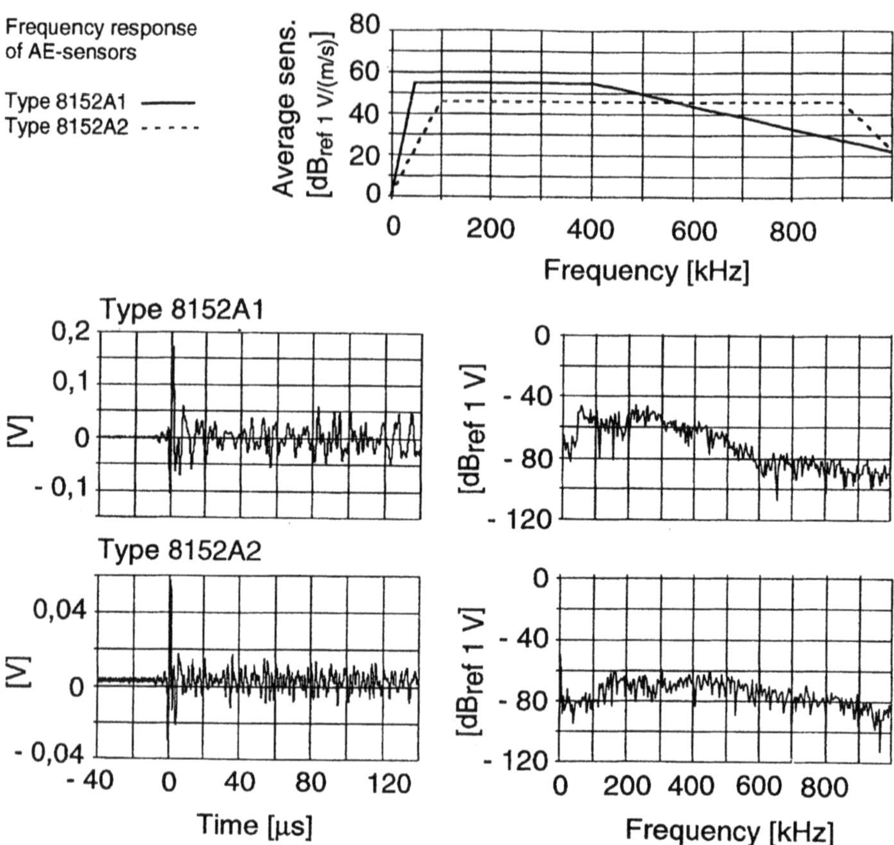

Figure 10.6 Recordings of AE sensor outputs (Courtesy of Kistler) AE signal (left) and its frequency spectrum (right) when breaking a 0,5mm, 2H pencil lead on a steel plate 200·200·30mm (Test after Nielson and Hsu). Sensors and point of excitation were on the same face, 50mm apart. Sensor type 8152A1 is optimized for a frequency range of 50...400kHz and a high sensitivity, while type 8152A2 is designed for a very wide frequency response of 100...900kHz

same block and their outputs are compared when AE is induced by breaking a lead mine at a point equidistant of the two sensors (White Noise Continuous Sweep [ASTM 1986: ASTM E976-84, Standard Guide for Determining the Reproducibility of AE Sensor Response], also known as the *face-to-face technique.*)

In process monitoring, steel is the most commonly encountered material and therefore, only the steel test block calibration is briefly shown here to serve as an example. The fundamental problems in calibrating an AE sensor are:

- AE sensors are usually sensitive in only one direction (normal to their coupling face or base) while a point on the surface of a test block may move in any

direction (e.g. normal to the sensitive axis of the sensor), because AE may be in the form of surface waves, bulk longitudinal waves and shear waves,
- the magnitude of displacement will be altered by the presence of the sensor,
- the sensor does not only contact the test block in a point but covers a certain surface area in which the displacement is not only a function of time but of location within the contact area, too.

Frequently a steel test block approximating a semi-infinite half space of steel is used. Because of the relatively large size of the test block the mechanical impedance at the surface is essential only depending on the material and not on the size of the block within the usual frequency range of interest to AE.

A well known method is the test after Nielsen and Hsu (ASTM Standard E976-84) [ASTM 1986] which uses a steel plate of 262x262x40mm. The AE signal is generated by breaking a 0,5mm pencil mine of hardness 2H on the large face of the steel block at a distance of 50mm from the sensor that is mounted on the same surface. Fig. 10.6 shows graphs obtained when calibrating an AE sensor by this method, but using a slightly smaller steel plate.

More information on calibrating AE sensors used in nondestructive testing can be found in [ISO 1998 and 1999; Miller and McIntire 1987].

11 Amplifiers for Piezoelectric Sensors

11.1 Quantity and Units of Measurement

The output of piezoelectric sensors is electric charge. The unit for electric charge in the SI (Système International) is the "Coulomb" (C), defined as $1\,C = 1\,A \cdot s$. The quantity of electric charge yielded by piezoelectric sensors is usually in the order of pC only, which is the reason why in practice the "pC" has become the commonly accepted working unit.

Electric charge as an electric quantity is generally not so well known and understood as e.g. voltage and current. Electric charge can be thought of as "quantity of electricity" with 1 Coulomb being the quantity of electricity transported by a current of 1 Ampère during 1 second. The examples given in Table 11.1 may help to visualize the orders of magnitude.

Table 11.1 Examples of electric charge

Elementary charge (of an electron or a proton)	0,000 000 000 000 000 000 16	C = 160 zC
Threshold of a modern charge amplifier	0,000 000 000 000 001	C = 1 fC
Charge yielded by an X-quartz plate loaded with 10 N (≈1 kgf)	0,000 000 000 023	C = 23 pC
Charge yielded by a load washer loaded with 1 MN (≈100 t)	0,000 002	C = 2 μC
Charge in a single lightning stroke	20	C = 20 C
Charge in a typical car battery	200 000	C = 200 kC

11.2 Principles of Measuring Electric Charge

11.2.1 Piezoelectric Sensor as an Active Capacitor

Piezoelectric sensors belong to the small group of active sensors (see chapter 1) and therefore, in principle, they do not require an external source of power to produce an output signal. This can be demonstrated with the classic gold-foil electrometer (Fig. 11.1). The angle of spread between the two foils α is proportional to the voltage U_{out} which in turn is proportional to the electric charge Q yielded by the piezoelectric element and, therefore, is proportional to the acting force.

If we take for the piezoelectric element a single X-cut quartz plate (sensitivity d_{11}=2,3pC/N), assume a force of 1kN and a sensor capacitance of 10pF, we get for U_{out}=(2,3pC/N·1kN)/10pF=230V. This open-loop voltage can reach several thousand Volt in some sensors and may even lead to a spark over at the open connector – which is the working principle of the piezoelectric gas lighter. In systems using an electrometer-type amplifier (see 11.2), care must be taken to keep the maximum voltage within acceptable limits. With charge amplifiers, this voltage is kept to zero (see 11.3).

Electrically speaking a piezoelectric sensor is a capacitor with the piezoelectric element acting as dielectric. Because here the dielectric exhibits a piezoelectric effect, *a piezoelectric sensor can be considered as an active capacitor that charges itself when mechanically loaded.*

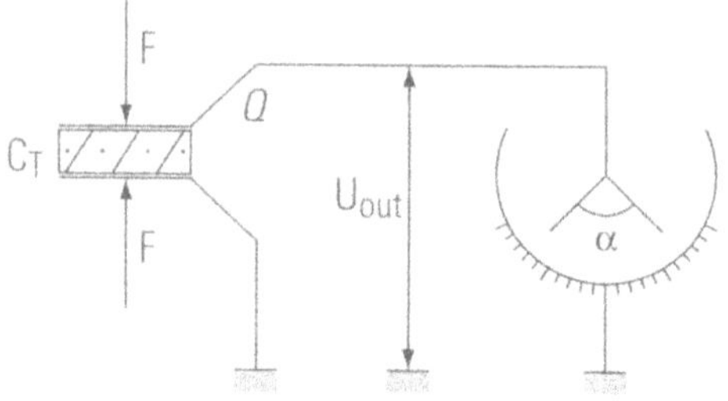

F	=	force acting on piezoelectric element (quartz)
C_T	=	capacitance of transducer
Q	=	charge yielded by quartz
U_{out}	=	output voltage
α	=	opening angle between the 2 goldfoils
d_{11}	=	sensitivity of an x-cut quartz (longitudinal effect)

Figure 11.1 Gold foil electrometer

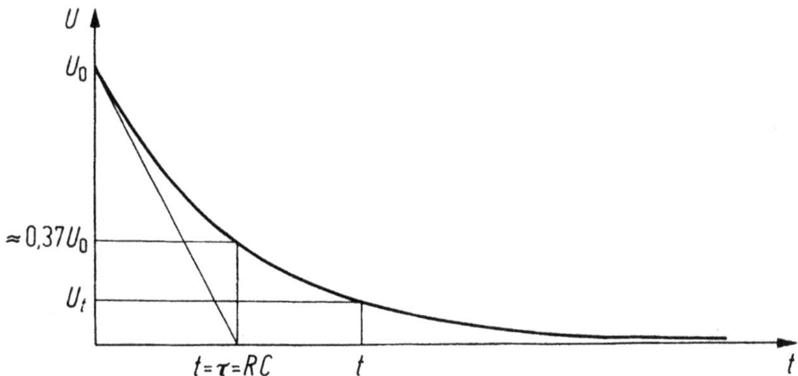

Figure 11.2 Discharge curve of a capacitor with a finite insulation resistance

11.2.2 Discharge of a Capacitor, Time Constant, Insulation Resistance

Electric charge is usually stored in capacitors whereby the voltage building up over the capacitor according to the relation $U=Q/C$ can be more easily be further processed as a measuring signal.

A capacitor charged by an electric charge Q to a voltage U would keep that voltage indefinitely if the capacitor had an infinitely high insulation resistance. In practice insulation resistance is not infinitely high and therefore, the voltage U drops eventually to zero. The reason for this is the discharge current $I=U/R$ which flows over the finite resistance R. As the voltage drops continuously, the current will also diminish. Therefore the voltage across the capacitor follows an exponential function (Fig. 11.2) with

$$U(t)=U_0\,\mathrm{e}^{-\frac{t}{\tau}}. \tag{11.1}$$

If the initial discharge current $I_0=U_0/R$ would continue to flow, the discharge would be finished after $t=\tau=R\cdot C$. This time τ is called "time constant" of the capacitor. As can easily be shown such a discharge line would correspond to the tangent on the real discharge curve at its beginning. In reality the discharge lasts indefinitely as the discharge curve approaches the zero line asymptotically.

After the time $t=\tau$ the voltage has dropped to

$$U_\tau=U_0\,\mathrm{e}^{-1}\approx 0{,}37\,U_0 \tag{11.2}$$

The time constant τ is a very useful parameter because it defines the initial tangent on the discharge curve. In the range $0<t<0{,}1\,\tau$ it is for all practical purposes a sufficient approximation of the curve and allows to determine easily the maximum duration of a measurement within a given limit of error. For a

Table 11.2 Typical insulation resistance of components used in electrometer and charge amplifiers as well as in sensors, cables and connectors. The values given apply to a insulation material that is clean and dry.

Material	Application	Insulation Resistance in TΩ at 20°C
PTFE	Cables, supports, connectors	>1000
Ceramics, siliconized	Switches, supports	> 100
Allyl resin	Switches, supports, connectors	> 100
Glass, highly insulating and siliconized	Reed relays	>1000
Glass, ordinary	(for comparison)	> 1
Magnesium-, silicon-, beryllium-oxid	Cables for high temperatures	> 10
Glass-epoxy	Support plates for printed circuits	> 1
Acrylic glass	Rotors for stepping switches	> 1

maximum error of e.g. 2%, the measuring time must not exceed 0,02 τ, i.e. 2% of τ. If required measuring time is e.g. 100 s, the time constant τ must be at least 5000 s.

The time constant is defined as $\tau = R \cdot C$. The capacitance C of the sensor and the cable is usually given by the design of the sensor and the type of cable. Only the range capacitor in the amplifier can be chosen within certain limits. In electrometer amplifiers, the range capacitor is chosen as to keep the voltage within the working limits of the amplifier, which is usually ±10 V. Although the time constant could be lengthened by increasing the capacitance, the resulting lower voltage limits this approach, because the signal-to-noise ratio would become insufficient. In charge amplifiers, other effects such as drift are usually limiting the length of measuring time (see 11.5.5). The second parameter determining the time constant is the resistance R which is essentially given by the insulation resistance of the materials used.

For obtaining the longest possible time constants, materials with the highest possible insulation resistance must be used. While in general electrical applications values over over 100 MΩ are already considered as sufficient or even excellent insulation, piezoelectric systems require insulation resistance of at least 1 TΩ, better 10 TΩ for achieving a sufficiently lower frequency limit and for allowing quasistatic measuring. Only for purely dynamic measurements and with sensors having piezoelectric ceramics as transduction elements, resistance in the order of GΩ are sufficient.

The insulation resistance of piezoelectric materials are mentioned in chapter 3. Table 11.2 shows the insulation resistance of the materials commonly used in sensors, cables and amplifiers.

11.2.3
Lower Frequency Limit of an *RC*-Circuit

The time constant determines also the lower frequency limit. Only with an infinitely long time constant, meaning the lower frequency limit is exactly zero, would it be possible to make true static measurements. This is impossible because the required infinitely high insulation resistance does not exist. This is the only reason why true static measurements are not possible with piezoelectric sensors. Only "quasistatic" measurements can be made (see 11.5.5).

For a sinusoidal measuring signal, the lower frequency limit is

$$f_u = \frac{1}{2\pi\tau}. \tag{11.3}$$

At this frequency limit, the amplitude of a sinusoidal signal is reduced to $1/\sqrt{2}$, i.e. to $\approx 0{,}707$. This reduction of the amplitude by about 30% corresponds to a damping of 3 dB (the Decibel being defined as $1\,\text{dB} = 20 \log U_1/U_2$). At the frequency limit not only the amplitude is reduced by 3 dB but there is also a phase shift of 45° between input and output signal (Fig. 11.3).

When measuring the measurands force, strain, pressure and acceleration considered here, amplitude errors of 30% and phase shifts of 45° are obviously not acceptable. The frequency limit given by the determining time constant must by lower by a factor of 5 to 10 than the lowest needed measuring frequency, depending on the amplitude error and phase shift allowed. Fig. 11.4 shows the amplitude error and phase shift as a function of the ratio measuring frequency to lower frequency limit of the circuit being a simple *RC* circuit with a roll off of 6 dB/octave below the frequency limit.

Signals from measurements seldom are purely sinusoidal and therefore the lowest frequency which is contained in the signal (e.g. the lowest frequency found by a FFT-analysis) must be taken as the decisive criteria for the required lower limit of the frequency range.

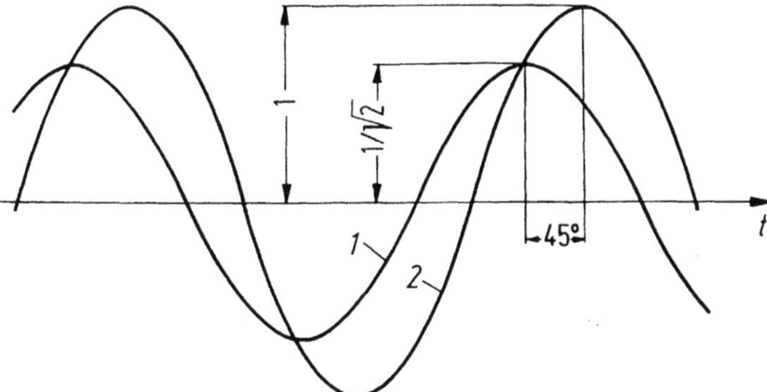

Figure 11.3 Amplitude error and phase shift of the output curve (*1*) at the limit frequency of a simple *RC* high-pass filter against the true curve (*2*)

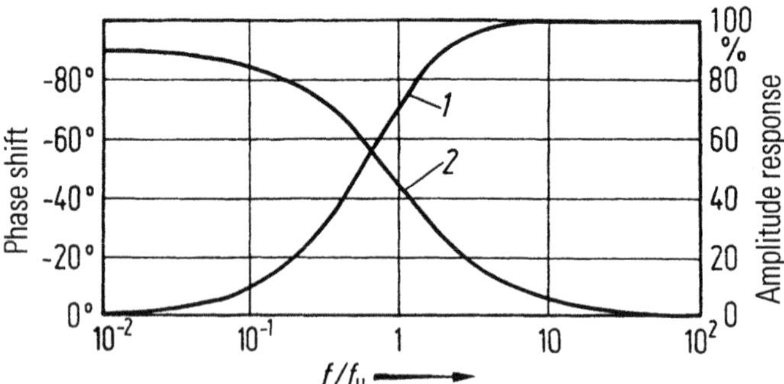

Figure 11.4 Amplitude response (*1*) and phase shift (*2*) of a simple *RC* high-pass filter as a function of the ratio of measuring frequency to limit frequency (ignoring the phase inversion of 180° inherent of most charge amplifier)

In most cases there are several successive time constants acting in a measuring system, e.g. in a charge amplifier the time constant of built-in filters and the time constant of the capacitive feedback circuit of the amplifier. For the measurement the highest of the various lower frequency limits is determinant. If several such frequency limits are close to each other, they will provoke a steeper roll off than a simple *RC* circuit.

11.2.4 Zero Point of a Measurement

A particular feature of the piezoelectric measuring system is the possibility of freely choosing the zero point for a measurement. Put even more correctly one can say that a piezoelectric measuring system inherently has no zero reference point. Zero must be defined intentionally before each measurement, either manually by resetting the amplifier or automatically by appropriate circuitry. Only of a piezoelectric transduction element which is not loaded (nor mounted and preloaded nor loaded by any other acting force) one could speak about an "inherent" zero point. Such a condition, however, does not correspond to any real practical situation and has therefore no useful meaning.

The principle of selecting and defining the zero point for a measurement can best be illustrated with a force sensor. If the force sensor is loaded with a constant force such as a weight placed on the sensor, the voltage $U=Q/C$ (Q being the electric charge yielded by the sensor under the mechanical loading, C the capacitance of the sensor) appearing at the sensor output will eventually drop to zero because of the discharge across the not infinitely high insulation resistance. When a new force is acting on the sensor now, the corresponding voltage will be referring to that "new" zero point.

Instead of waiting for the voltage to drop to zero after the initial loading of the sensor, the sensor output can simply be short-circuited (if the magnitude of the first loading is of no interest, then the short-circuiting can take place even before the initial loading!). With the zero so defined, further acting forces will be measured relative to that zero point. The key point is that it does not matter how large or small this initial load is. Starting from the zero point so established we are free to choose the measuring range on the amplifier depending on the force to be measured now. If we have e.g. an initial load of 100 kN, we can reset electrically to zero by just short-circuiting the sensor output or by resetting the amplifier and then measure e.g. a force of a few N only with a correspondingly selected range, obtaining an optimal signal-to-noise ratio.

This feature is especially useful in multicomponent force sensors which have to be mounted always under a high preload. Since zero is set before each measurement, the already existing preload has no influence on or meaning for the measurement about to be made. This type of establishing the zero point is nothing else than "taring" as it is known from the field of weighing, except that the possible range of taring is several orders of magnitude bigger.

Although it is inherently not possible to measure statically over a longer period of time, the property just described allows to come back to the sensor after an indefinitely long time, connect an amplifier, and measure the still existing preload by taking the preload off, resulting now in a negative force step as output.

11.3 Ideal Electrometer Amplifier

The electrometer amplifier is an amplifier with a very high insulation resistance at its input, ideally an infinitely high insulation resistance. Therefore the input circuit – sensor and cable – is not loaded and the input voltage will be maintained. The gain is usually 1, i.e. there is no change in the voltage. The electrometer amplifier acts as an impedance converter. The voltage $U=Q/C$ appearing at the sensor output is converted into an equal voltage across a very low resistance at the amplifier output. This output voltage can be measured with the usual instruments such as a voltmeter, oscilloscope, A/D converter, etc.

Fig. 11.5 shows a measuring chain with an electrometer amplifier. The voltage at the amplifier input is determined by the electric charge yielded by the sensor and the total capacitance in the input circuit. In order to draw off a minimum of electric charge from the input circuit the insulation resistance of the amplifier

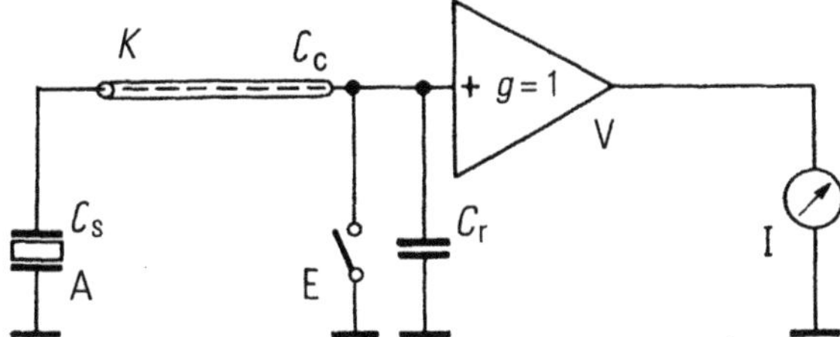

Figure 11.5 Measuring chain with electrometer amplifier. A sensor, K cable, V amplifier, E reset ("earthing" or "grounding") switch, I read-out instrument (e.g. voltmeter)

should ideally be several orders of magnitude higher than the insulation resistance of sensor and cable. Originally special electrometer tubes were used which have been replaced by the modern field-effect transistors (MOS-FET and J-FET). In an ideal electrometer amplifier, the insulation resistance in the input circuit would be infinite, the input stage (electrometer or transistor) would be free of any leakage current, and true static measurements would be possible.

For adapting the measuring range to a specific measuring task, i.e. to limit the voltage to the maximum voltage accepted by the amplifier, one or several switchable range capacitors C_r are built into the amplifier. The total capacitance within the input circuit is composed of the sensor capacitance C_s, the cable capacitance C_c and the switched-in range capacitance C_r. The voltage appearing in the input circuit is

$$U_e = \frac{Q}{C_s + C_c + C_r}. \tag{11.4}$$

Depending on the electric charge to be measured, C_r is chosen to keep the voltage U_i within the working range of the amplifier. Electrometer tubes as well as field effect transistors usually accept input voltages from $\pm 1\,\mathrm{V}$ up to $\pm 10\,\mathrm{V}$, depending on the type of device.

An electrometer amplifier usually has a so-called "grounding switch" with which the input circuit can be short-circuited to set the zero point for the next measurement (see 11.5.3). Moreover – with this grounding switch closed – the amplifier input is protected against voltage peaks which may occur when e.g. connecting a statically charged cable.

11.4 Real Electrometer Amplifier

The output voltage of the electrometer amplifier depends on the gain v of the amplifier and will be

$$U_o = v\,U_i = \frac{Q}{C_s + C_c + C_r}. \tag{11.5}$$

The gain v is usually 1 and therefore $U_o = U_i$.

From (11.5) can be seen that the output voltage of the amplifier depends not only on the selected range capacitance C_r but to a large extent also on the sensor capacitance C_s and the cable capacitance C_c. The sensor capacitance is in the order of some ten to several hundred pF and usually within quite close tolerance of the value specified for a given type of sensor. The cable capacitance is directly dependent on the cable length and typically is about 70 pF/m.

In order to determine the transfer factor from the electric charge Q – which is proportional to the measurand – to the output voltage U_o of the amplifier, it is necessary to either measure and precisely and enter their values into (11.5) or to calibrate the complete measuring chain. After exchanging the sensor and certainly after changing the cable, especially when it is of a different length, the capacitance must be measured again and the sensitivity of the system be recalculated, or the whole system must be recalibrated.

A further inconvenience is that the output voltage is not proportional to the capacitance of the range capacitor which leads to "odd" values for the transfer factor or "sensitivity" of the system. A voltage amplifier stage with adjustable gain is sometimes added to the electrometer amplifier in order to arrive at "even" values for the sensitivity. However an amplifier so adjusted is only correct for the calibrated setup and the particular range capacitor selected.

The total capacitance in the input circuit determines, together with the total insulation resistance, the time constant of the measuring system. The insulation resistance corresponds to the values of the insulation resistance of the sensor, of the cable and of the amplifier input stage switched in parallel. Therefore, the measuring chain behaves exactly like a sensor with an open output.

A measuring chain with an electrometer amplifier will always go to zero (or to a value determined by the grid current of the electrometer tube), if left alone for a sufficient length of time. The time constant of the system, defined by the total input capacitance and the insulation resistance, determines the lower frequency limit. The effective time constant can be shortened by switching in a resistor parallel to the amplifier input. This increases the lower frequency limit (see 11.5.2). For quick resetting or taring to determine the zero point, a grounding switch, often called a reset switch, can be added across the input of the amplifier (Fig. 11.5 and chapter 11.5.3).

For measuring slow phenomena, i.e. in a quasistatic mode, a time constant of maximum length is needed. As shown in 11.2.2 this means that the insulation resistance in the input circuit should be as high as possible. In sensors with transduction elements of quartz and at the input of electrometer tubes, insulation values of over 10TΩ can easily be maintained. Cables and connectors with PTFE as insulation material easily achieve values of over 10TΩ, when new and clean. After prolonged use without proper care and maintenance, especially for cleanliness, their insulation can drop to less than 1TΩ. Measuring quasistatically with such low insulation values becomes difficult or even impossible when using electrometer amplifiers, particularly when small signals have to be measured over a certain length of time.

This problem in the earlier days of piezoelectric measuring has led to the view that by principle it was impossible to measure slow or quasistatic phenomena with piezoelectric sensors and that maintaining the cable insulation was a major obstacle for the user. This argument which was certainly valid up to the 1960's has developed into a non-objective prejudice against the piezoelectric system which persists still today, curiously enough particularly in the USA and in Japan. With the introduction of charge amplifiers in the late 1950's, cable insulation and capacitance have become much less critical parameters in quasistatic measurements. As shown in chapters 11.4 and 11.5.5, such measurements can be made reliably with charge amplifiers, for which reason the piezoelectric sensor have found a multitude of applications where slow or quasistatic phenomena have to be measured or monitored.

Further difficulties in electrometer amplifiers with electrometer tubes as input stage is the poor linearity of such vacuum tubes (mostly pentodes) and their sensitivity to shock and vibration which produces the so-called "microphone effect". The poor linearity of the electrometer tube is found in the relation between input and output signal, making calibration and signal evaluation difficult. The microphone effect severely limits the application of such amplifiers in rough environments, especially in industrial monitoring, where vibration and noise are often present. For all these reasons electrometer amplifiers have been superseded almost completely by charge amplifiers.

An important exception are the sensors with built-in amplifier where the electrometer amplifier continues to be the preferred choice. Of course, electrometer tubes are not suitable and MOS-FET and J-FET are used. Obviously the capacitance of the input circuit is constant and fixed once and for all when the sensor is built. Also the insulation resistance of the input circuit poses no problem because such sensors including the electronics are usually hermetically sealed.

These sensors are used mainly for measuring vibration which is by its nature a dynamic and not a quasistatic measurand. Force sensors are also available with built-in amplifiers for dynamic applications. Therefore a long time constant is not required. Usually, there is even a resistor added in parallel to the input stage to obtain a shorter time constant and a correspondingly higher lower frequency limit. This makes a grounding switch superfluous which would be difficult to implement anyhow.

The operating temperature range of sensors with built-in amplifiers is naturally limited by the temperature range acceptable to the electronic components. With standard electronic components, the operating temperature range usually covers about –40...125°C. With special electronic components, the upper limit can be pushed to about 165°C.

Sensors with built-in amplifiers are marketed and known under several trade names, such as DeltaTron® (Brüel & Kjær), ICP® (PCB), ISOTRON® (Endevco) and Piezotron® (Kistler). They all work on the 2-wire principle, i.e. they are powered by a constant current (usually 4 mA). The electrometer circuitry built into the sensor behaves like a resistor whose resistance varies as a function of the measurand. A precise resistor (usually 100 Ω) is inserted in series with the sensor and the voltage drop across it also becomes a function of the measurand.

This 2-wire system works with ordinary cables (no need for coaxial cables or shielding) and the cable length can be substantial (up to several hundred meters) without degradation of the signal or high frequency cutoff. Cable movement, e.g. caused by vibration, does not affect signal quality. Also, a harsh environment (humidity, dust, dirt) is not critical for operation.

Detailed information on the variety of electronics available for driving sensors with built-in electronics is published by the manufacturers and can also be found in [Mahr and Gautschi 1982].

11.5 Ideal Charge Amplifier

The term "charge amplifier" – widely used for many years – is a misnomer because a charge amplifier does not amplify electric charge but rather converts electric charge into a proportional electric voltage. The term "charge meter" recently introduced by one manufacturer (Kistler) may be a viable alternative because many charge amplifiers have now displays and can, therefore, be likened to e.g. voltmeters. For the sake of simplicity, the term "charge amplifier" is retained here.

The charge amplifier has largely contributed to the rapid spread and acceptance of piezoelectric sensors in ever wider fields of applications, because it does away with the drawbacks of electrometer amplifiers.

The principle of the charge amplifier was first described by W. P. Kistler in 1950 [Kistler 1950]. Basically a charge amplifier is just an inverting DC amplifier with an internal gain as high as possible, and an insulation resistance at its input as high as possible. This amplifier has a highly insulating capacitor as capacitive feedback which brings the voltage at the input to zero while maintaining the high insulation resistance.

The operating principle of a charge amplifier can be explained as follows (see Fig. 11.6): The output voltage U_o is related to the input voltage U_i by

$$U_o = -v_i\, U_i, \tag{11.6}$$

where v_i is the inner gain of the operational amplifier. The voltage difference over the feedback capacitor is

$$U_C = U_o - U_i = U_o - \frac{U_o}{-v_i} = \left(1 + \frac{1}{v_i}\right) U_o. \tag{11.7}$$

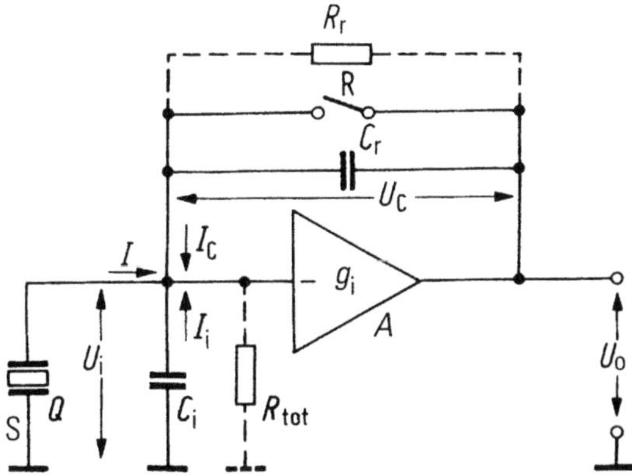

Figure 11.6 Measuring chain with charge amplifier. A sensor, V amplifier, R_{tot} insulation resistance at the input, R reset switch, C_c capacitance at the input

At the input of the amplifier, according to Kirchhoff's law, the sum of all currents is zero. Assuming an ideal charge amplifier, its input stage is free of leakage current, i.e. no current flows into the amplifier input. Therefore

$$I+I_{\mathrm{C}}+I_{\mathrm{i}}=0, \tag{11.8}$$

whereby

$$I=\frac{\mathrm{d}Q}{\mathrm{d}t}, \tag{11.9}$$

$$I_{\mathrm{C}}=C_{\mathrm{r}}\frac{\mathrm{d}U_{\mathrm{C}}}{\mathrm{d}t}=\left(1+\frac{1}{v_{\mathrm{i}}}\right)C_{\mathrm{r}}\frac{\mathrm{d}U_{\mathrm{a}}}{\mathrm{d}t} \tag{11.10}$$

and

$$I_{\mathrm{i}}=C_{\mathrm{i}}\frac{\mathrm{d}U_{\mathrm{i}}}{\mathrm{d}t}=\frac{1}{v_{\mathrm{i}}}C_{\mathrm{i}}\frac{\mathrm{d}U_{\mathrm{o}}}{\mathrm{d}t}. \tag{11.11}$$

The total capacitance C_{c} in the input circuit is the sum of the capacitance of the sensor, the cable and at the amplifier input. It follows

$$\frac{\mathrm{d}Q}{\mathrm{d}t}=-\left(1+\frac{1}{v_{\mathrm{i}}}\right)C_{\mathrm{r}}\frac{\mathrm{d}U_{\mathrm{o}}}{\mathrm{d}t}-\frac{1}{v_{\mathrm{i}}}C_{\mathrm{i}}\frac{\mathrm{d}U_{\mathrm{o}}}{\mathrm{d}t}. \tag{11.12}$$

After integrating and ignoring the constant of integration (this corresponds to resetting of the amplifier described in 11.2.4) we get for the output voltage

$$U_{\mathrm{o}}=-\frac{Q}{\left(1+\frac{1}{v_{\mathrm{i}}}\right)C_{\mathrm{r}}+\frac{1}{v_{\mathrm{i}}}C_{\mathrm{i}}}. \tag{11.13}$$

The inner gain v_{i} is assumed to be infinitely high in the ideal case here and equation (11.13) simplifies to

$$U_{\mathrm{o}}=-\frac{Q}{C_{\mathrm{r}}}. \tag{11.14}$$

It is evident that the output voltage U_{o} is now directly proportional to the electric charge yielded by the sensor, because the range capacitor C_{r} is constant for a given measurement. Above all, the capacitance in the input circuit, i.e. sensor and cable capacitance have no influence on the measuring result anymore. Since the electric charge yielded by the sensor is also proportional to the measurand, it follows that the output voltage of a charge amplifier is directly proportional to the measurand.

In the ideal case with an infinitely high gain the input voltage $U_{\mathrm{o}}= U_{\mathrm{o}}/v_{\mathrm{i}}=0$, i.e. there is no voltage across the sensor and the cable! The respective "capacitors" can not be charged and will have no influence. Furthermore, because the input voltage is zero, no current can flow across the resistance R_{tot} in the input circuit. In the ideal case, even the insulation resistance of sensor and cable would not play any role. Therefore, at the amplifier input only the the current originating from the electric polarization charge in the sensor and the charging current of the range capacitor (feedback capacitor) are present. Kirchhoff's law says that these two currents are equal but of opposite polarity at all time. Therefore the electric charge from the sensor and the electric charge in the range capacitor are equal but of opposite polarity. *It seems as if the electric charge yielded by the sensor flowed directly into the range capacitor.* Or put still in an other way: *The charge amplifier continuously compensates the electric charge yielded by the sensor with an equal charge of opposite polarity in its range capacitor.* With the input voltage in the ideal case at zero, it further follows that $U_o = U_C$.

For fixing the zero point before a measurement a switch (so-called "reset switch") is added in parallel to the range capacitor, allowing the latter to be discharged before a measurement and thus defining the zero point at that instant (see 11.5.5).

The quasistatic behavior of the amplifier is now determined only by the time constant $\tau = R_r \cdot C_r$ in the capacitive feedback circuit, i.e. by the product of the capacitance of the range capacitor and the insulation resistance in the feedback circuit (essentially the insulation resistance of the range capacitor itself). This insulation resistance of the range capacitor would be infinitely high in the ideal case, making possible true static measurements, i.e. offering true "DC mode". For obtaining an "AC mode" response, additional resistors can be switched in parallel with the range capacitor, reducing the capacitance which decides the time constant (see 11.5.5.1).

The advantages of the ideal charge amplifier emerge clearly from the characteristics just described:

- no voltage develops across the sensor and cable capacitances, which means that the insulation resistance in the input circuit is not critical.
- the capacitance in the input circuit (sensor and cable) has no influence on the output voltage.
- the output voltage is directly proportional to the electric charge yielded by the sensor and hence proportional to the measurand, too.
- the output voltage is inversely proportional to the feedback capacitor, allowing to easily obtain conveniently stepped measuring ranges by simply using capacitors of appropriate values (for that reason, these feedback capacitors are often called "range capacitor").

All these characteristics are essentially found in a real charge amplifier, too.

11.6 Real Charge Amplifier

In practice, the decisive parameters in 11.4 will only come more or less close to their idealized values. The highest gain found in operational amplifiers is only about 50 000 to 100 000; the insulation resistance at the amplifier input as well as for the range capacitor reaches at best about 100 TΩ; and there are no input transistors which are completely free of leakage current (about 1 fA is already an excellent value). The ramifications of these "imperfections" for the behavior and therefore the practical use of real charge amplifiers is described in detail in the following chapters and their implications for the measurement practice are discussed.

11.6.1 Sensitivity Setting, Scale and Measuring Range

Piezoelectric sensors are usually calibrated by the manufacturer and their sensitivity is expressed in pC/M.U. (M.U. = mechanical unit), e.g. pC/N, pC/με, pC/bar, pC/g, etc. Occasionally, sensor sensitivity expressed in V/M.U. can be found. Such a practice is discouraged because it is quite a useless information unless the exact value of the total capacitance of sensor, cable and charge measuring device that were used during calibration is indicated as well. Therefore, the charge sensitivity should always be given, because electric charge – and not voltage – is the primary output of piezoelectric sensors. The voltage sensitivity can easily be found at any time by simply dividing the charge sensitivity by the capacitance present in the given application (sensor and cable capacitance). Additionally, any range capacitor used in an electrometer amplifier has to to be taken into account, too.

Charge amplifiers invert the sign of the input signal (see formula 11.14), i.e. they give a negative output voltage in response to a positive electric charge at their input and vice versa. Because a positive output voltage in response to a positive measurand acting on the sensor is desirable, the transduction elements are inserted into a piezoelectric sensor in such a way as to give a negative charge output in response to a positive measurand. This practice has been followed by practically all manufacturers for decades and therefore nearly all piezoelectric sensors have a "negative" sensitivity, e.g. –4,3 pC/N or –9,6 pC/bar. Well-designed microprocessor-controlled charge amplifiers allow to enter the negative sign of the sensors sensitivity and there is no ambiguity about the sign of the output signal. While in force and pressure measurements, the correct sign usually can be checked quite easily, this is often not so in vibration measurements. Using the wrong sign can lead to errors, e.g. in terms of phase shift.

In general, the sensor sensitivity is not an "even" number but can be any value (e.g. –3,87 pC/N, –11,8 pC/bar, etc.). A charge amplifier offering only a fixed sensitivity (e.g. 1 V/pC) will not give an "even" scale at its output, which makes signal processing inconvenient. Therefore, most charge amplifiers have an adjustable sensitivity ("gain" in terms of V/pC) which in its most simple form can be an adjustable range capacitor or a trimming potentiometer for adjusting the gain of a voltage amplifier following the charge amplifier. This is satisfactory where a measuring chain is set up once and for all, such as in industrial monitoring systems. Often, "odd" values of the sensitivity are no longer of much significance when signal processing is done in a microprocessor or a computer, because there the "correction factor" can easily be programmed in. Another approach is to use a so-called "calibration plug" which contains the individual calibration data of the sensor and which by simply plugging it into the amplifier makes the required adjustment.

For laboratory-type measuring tasks amplifiers that allow direct setting of value of the sensitivity of the sensor connected are the preferred choice. Such amplifiers have e.g. a 10-turn precision potentiometer for setting the value and a switch for setting the decimal place. However, charge amplifiers that use a microprocessor are now the preferred choice, making the setting for a convenient and "even" scale much easier with their menu-guided setup.

The total sensitivity S_{tot} is obtained by multiplying the sensor sensitivity S_s and the amplifier sensitivity S_a, i.e. $S_{tot} = S_s \cdot S_a$. For e.g. $S_s = -7{,}68\,\text{pC/N}$ and $S_a = -0{,}01302\,\text{V/pC}$, the overall sensitivity becomes $S_{tot} = -7{,}68\,\text{pC/N} \cdot (-0{,}01302\,\text{V/pC}) = 0{,}1\,\text{V/N}$.

Although this is quite a common way to calculate it has the disadvantage that for determining the evaluation scale for connected indicators, oscilloscopes, recorder or even A/D converters, further calculation is needed. If in the example above, a recorder with a sensitivity of 1 V/cm is used, the scale of the record has to be calculated as (1 V/cm)/(0,1 V/N) = 10 N/cm. Far more elegant and simple is to use the reciprocal value, i.e. to indicate how many mechanical units correspond to 1 Volt of the output voltage of the amplifier. In the example above, this works out to 1/(0,1 V/N) = 10 N/V. For the recording we then get 10 N/V · 1 V/cm = 10 N/cm.

This approach has the obvious advantage that the scale "M.U./V" selected on the charge amplifier becomes directly the scale of the recording if for the latter, 1 V/cm or 1 V/div is selected.

The possible measuring range can be obtained by multiplying the scale with the maximum voltage available at the amplifier output. The most common output voltage ranges are either ±10 V (especially for laboratory-type amplifiers) and also ±5 V, ±2 V or ±1 V (especially for industrial-type amplifiers). In the laboratory a choice of several measuring ranges is needed. Up to a certain degree, this can be obtained by adjusting the gain of the voltage amplifier following the charge amplifier stage. There are limits to that because of the signal-to-noise ratio which will be reduced. Therefore laboratory-type charge amplifiers have several range capacitors that can be selected.

Commonly used range stepping is by full decades or by the sequence 1, 2, 5, 10, 20 and so on. In noise and vibration work, often the stepping 1, 3, 10, 30,...is preferred when calculations are done in dB. With 10 dB representing a voltage ratio of $\sqrt{10} = 3{,}16$, the sequence 1, 3, 10, 30, ... corresponds closely to 10 dB steps, unless

Figure 11.7 Microprocessor-controlled charge amplifier with menu-guided setting of the parameters. Offers DC-mode and AC-mode operation (Courtesy of Kistler)

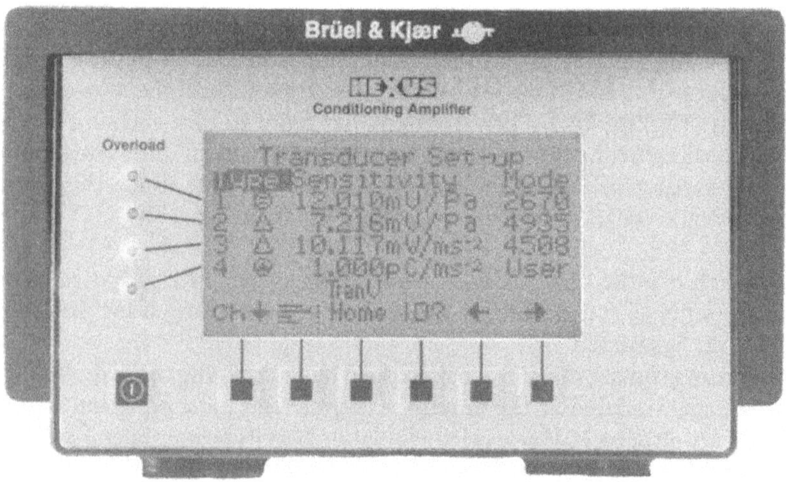

Figure 11.8 Microprocessor-controlled charge amplifier with 4 channels, especially designed for acoustic and vibration measuring, i. e. AC-mode operation (Courtesy of Brüel & Kjær)

true 10 dB steps are used directly. More recently, microprocessor-controlled charge amplifiers have been introduced, offering much greater flexibility in setting ranges and other operating parameters. Figs. 11.7 and 11.8 show examples of such amplifiers.

11.6.2 Lower Frequency Limit of a Charge Amplifier

Charge amplifiers can by divided into two groups: those with a clearly defined lower frequency limit (usually a few Hz) for AC-type (dynamic) measurements only and those with the lowest possible frequency response for DC-type (quasistatic) measurements.

The lower frequency limit depends primarily on the time constant of the charge amplifier, which is given by the values of the range capacitor and the resistance in parallel with it. The longest time constant is obtained when the only resistance in parallel with the capacitor is its own insulation resistance, which typically is more than 100 TΩ. This gives time constants of over 100 ks by calculation and therefore the possibility of measuring quasistatically. However, real charge amplifiers will usually not go to zero after a sufficiently long time – as would be the case with a true time constant – but eventually they always drift into positive or negative saturation if left alone. The main reason is the drift caused by the leakage current of the input stage which dominates the behavior of the charge amplifier and which is described in 11.5.5.1.

For dynamic measurements – AC-mode or AC-coupling – a resistor R_f is switched in parallel with the range capacitor C_r in order to obtain the time constant which will result in the required lower frequency limit. The resistance and capacitance from a simple *RC*-high-pass filter with a slope of 6 dB/octave. From the time constant

$$\tau = R_g C_g \tag{11.15}$$

results the lower frequency limit (–3 dB point)

$$f_u = \frac{1}{2\pi\tau} = \frac{1}{2\pi \, R_g C_g}. \tag{11.16}$$

That frequency limit must be sufficiently lower than the lowest frequency contained in the measuring signal (see 11.1.3) to keep the amplitude and phase errors within tolerated limits.

However the lower frequency limit can not be chosen just by these criteria alone because other conditions must be fulfilled to remain in an operational range where the amplifier will not go into saturation but always to zero if left alone (characteristic of AC-mode). The value of the time constant resistor R_f should be at least about 10 times lower than the insulation resistance of sensor and cable. This means e.g. about 100 GΩ for sensors with quartz elements and 1 GΩ for sensors with piezoelectric ceramic elements. Leakage currents and offset voltage in the input circuit may set further limits.

Charge amplifiers for dynamic measurements only usually have an appropriate resistor already soldered in while in charge amplifiers with quasistatic response, such resistors can be switched in with a “time constant switch”. Basically one resistor would be enough but because there is a factor of about 100 between the

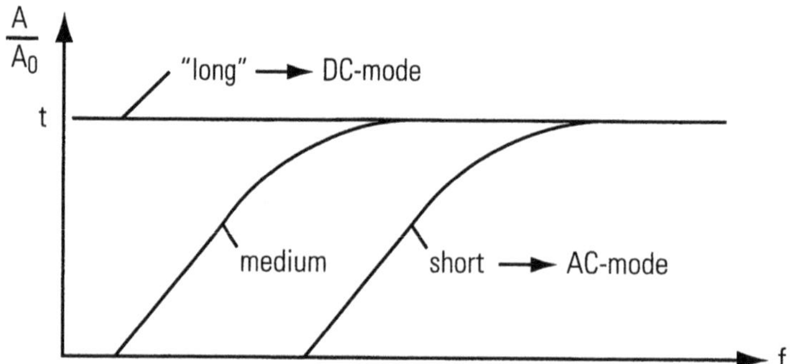

Figure 11.9 Low-frequency response of a charge amplifier in DC mode ("long" time constant) and AC mode ("medium" and "short" time constant). f frequency, A_0 input amplitude, A output amplitude

insulation resistance of quartz and piezoelectric ceramic sensors, commercially available charge amplifiers have in general two such resistors, i.e. two time constants, usually called "short" (mainly used with ceramic sensors) and "medium" (used with quartz sensors). A third "time constant", called "long" by some manufacturers, still found on certain charge amplifiers is not at all a time constant, because a charge amplifier left to itself will not go to zero when set to "long time constant", but simply saturate with time. Fig. 11.9 illustrates the lower frequency response of a charge amplifier in DC and AC mode. The particularities of the "long time constant" are explained in 11.5.5.

The short and medium time constants are determined just by the fixed time constant resistor and the range capacitor. In charge amplifiers with switchable ranges, the time constants and the respective lower frequency limit will depend on the selected range. In certain applications, especially in noise and vibration measuring this is often inconvenient because it may give inconsistent results when data analyzing equipment such as FFT and spectrum analyzers are used. Charge amplifiers have also been designed in which the lower limit frequency is kept constant by switching in appropriate resistors automatically to keep the time constant and therefore the lower frequency limit independent of the range selected. In practice this approach has its limits because resistors above 1 GΩ have rather large tolerances and are not very stable over time.

For critical applications it is preferable to insert a high-pass filter after the charge amplifier stage which can be made more precise and, if required, also with a steeper slope than 6 dB/octave. Of course the lower frequency limit of such a filter must be well above the lower frequency limit defined by the smallest range capacitor to be used and the feedback resistor.

The newest generation of microprocessor-controlled charge amplifiers often convert the signal into digital from. Signal processing is then done fully by digital circuits which offer a much greater freedom of filtering. The fully conditioned signal is the converted back to an analog output voltage for further handling, unless it can be fed directly into a fully digital processing or controlling system.

11.6.3 Resetting and Choosing the Zero of a Charge Amplifier

A charge amplifier operated on long time constant, i.e. with a very high resistor (over 100 GΩ) or no resistor at all in parallel with the range capacitor), must be provided with a reset switch, allowing to short-circuit the range capacitor for bringing the output voltage to zero (see 11.1.4). The switch must not only have a sufficiently high insulation resistance (preferably over 100 TΩ) but also not generate any electric charge when operated to minimize the so-called "operate jump" described in 11.1.4. Often reed relays sealed in special high-insulating glass that can be operated by an electromagnetic coil around them, or special semiconductors such as J-FETs are used. With medium and especially with short time constants, i.e. R_f=100 GΩ and 1 GΩ, and with charge amplifiers for dynamic measuring only, a reset switch is not necessary in principle because such amplifiers go quickly to zero when left to themselves (AC-mode), as described in 11.1.2.

11.6.4 Upper Frequency Limit of a Charge Amplifier

The upper frequency limit of a measuring chain is determined by that element which – with increasing frequency – starts first to cut (or increase) the amplitude and to cause a phase shift in the measuring signal. This can be the sensor, the cable or the charge amplifier itself. In most cases it is the sensor which sets the upper frequency limit. The natural frequency or the mounted resonant frequency of a piezoelectric sensor is in a range from about 1 kHz to over 500 kHz, mostly within 50 to 200 kHz. These values are often reduced by the mechanical conditions when mounting the sensor. At frequencies below about 20...30% of the natural frequency or the mounted resonant frequency of the sensor, amplitude and phase shift errors are usually within acceptable limits for most practical purposes (Fig. 9.5).

Cables rarely have an influence on the upper frequency limit except when they are very long (generally only above about 50 m), when very small measuring ranges are used in the charge amplifier, when older charge amplifiers with a rather low inner gain are used, and when measuring frequencies go over about 100 kHz.

The relation between inner gain v_i, feedback (or range) capacitance C_r and the cable capacitance C_c are explained in 11.4 and 11.5.6. The inner gain is frequency dependent, i.e. it drops with increasing frequency. Therefore the influence of the cable capacitance is no longer negligible as it will reduce the signal amplitude with increasing frequency in a way similar to a low-pass filter. Details on the cable influence are usually provided in the operating instructions for the charge amplifier by the manufacturer.

At frequencies above about 100 kHz the high-frequency impedance of the cable starts to play a role. It is no longer sufficient to just consider the cable capacitance alone. For an optimum transmission, sensors and amplifiers should actually be matched to the cable, but then problems with large range capacitors could arise at higher frequencies. In practice such high-frequency impedance matching is rarely done except perhaps when measuring in the MHz range, such as in shock wave studies or acoustic emission work.

The charge amplifier seldom limits the frequency range in most applications. Modern operational amplifiers with appropriate circuitry reach upper frequency limits of over 500kHz which is sufficient for practically all applications.

Certain limitations may be found only when the capacitance of the range capacitor is small compared with the cable capacitance or when very large range capacitors are used. With large range capacitors, the amplifier must deliver a large current (up to over 100mA) in a short time, i.e. it must have a high slew rate. These conditions are, however, rarely encountered with mechanical measurands because a high magnitude of a mechanical quantity is rarely combined with high frequency.

Should on the other hand the frequency range be limited at its upper end, it is easy to use low-pass filters in the amplifier, either already built-in and selectable by a switch or via the keyboard, or with plug-in filters.

11.6.5 Quasistatic Measuring, Stability and Drift

Despite a still widespread prejudice, piezoelectric measuring systems can be used perfectly well and reliably for measuring quasistatically, i.e. "like static" or in a "near-DC mode" over a certain length of time. A wealth of very successful applications, both in the laboratory and in industrial process control, are exploiting this capability.

There are a number of phenomena which determine the limit for quasistatic measuring. Theoretically, one limit is the time constant of the feedback circuit of the charge amplifier. In practice, the reference point of a charge amplifier will – in the quasistatic mode, i.e. in DC-mode or set to "long time constant" – nearly always show a drift, which can be in the positive or in the negative direction and which will drive the amplifier to go into saturation after a sufficiently long time. This phenomena – globally called "drift" – is dominant and results from a combination of several effects, defining the real behavior of the charge amplifier. Drift can have the following components:

- time constant in the feedback circuit,
- dielectric memory effect in the feedback (range) capacitor,
- leakage current in the input device (MOS-FET, J-FET, varactor diode),
- insufficient zero stability,
- leakage current in the input circuit caused by an offset voltage,
- offset voltage at the output due to poor insulation resistance in the input circuit and short time constant of the charge amplifier,
- "operate jump",
- equalization of electric charge generated by manipulating components of the measuring system (triboelectric effects),
- partial defect of the MOS-FET at the input.

11.6.5.1
Phenomenon "Drift"

Drift in charge amplifiers is still a little understood phenomenon. Drift – as defined by the ANSI/ISA-S37.1-1975 (R1982) standard [ISA 1982] – is "an undesired change in output over a period of time, which change is not a function of the measurand" (see 4.2.1.5). Under "normal" conditions – which means above all no or only small temperature changes, especially of the sensor – the output of a charge amplifier (in DC mode) will be a more or less linear drift of a few fC/s, which translates into a voltage drift at the amplifier output. It is not predictable, but limits for it can be specified. Good amplifiers (with MOS-FET input stage) have a drift that remains within about ±0,03 pC/s (Fig. 11.10). The amount of the resulting output voltage drift depends on the range selected on the amplifier. If e.g. the amplifier is set to a range of ±100 pC (which gives a sensitivity of 10 pC/V), the voltage drift will be within about ±3 mV/s. Note that drift can be "±", i.e. positive or negative! A charge amplifier – after having been switched to "operate" – will always drift into positive or negative saturation, if left alone for a sufficient time, and regardless whether a sensor is connected or not.

The practical implication of drift – which determines the "quasistatic" limit for a given measurement – can be illustrated with the example of measuring a static force, such as using a piezoelectric sensor like a balance. Let us assume that a quartz load washer (Fig. 6.2) is loaded with a weight of 100 kg or 10000 kg which corresponds to a force of about 1 kN or 100 kN on the sensor. The drift valid for such a load washer can easily be calculated from its sensitivity, which is roughly 4 pC/N. Drift will be within (±0,03 pC/s)/(4 pC/N) ≈ ±10 mN/s (the value for drift of ±0,03 pC/s is valid for most modern charge amplifiers with a MOS-FET at the input). It is very important to note that this drift does not depend on the measuring range selected in the charge amplifier.

Under laboratory conditions, drift of this magnitude or even smaller (when temperature variations can be kept within narrow limits and if the amplifier has been adjusted for minimal drift, see 11.5.5.5) can readily be obtained, making quasistatic measurements over quite long time spans possible.

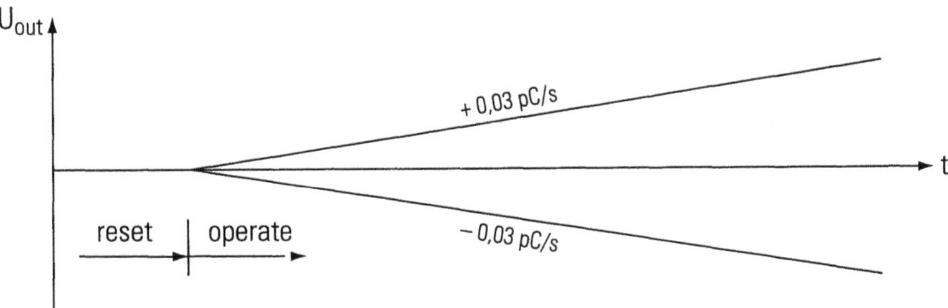

Figure 11.10 Drift in a charge amplifier in DC-mode

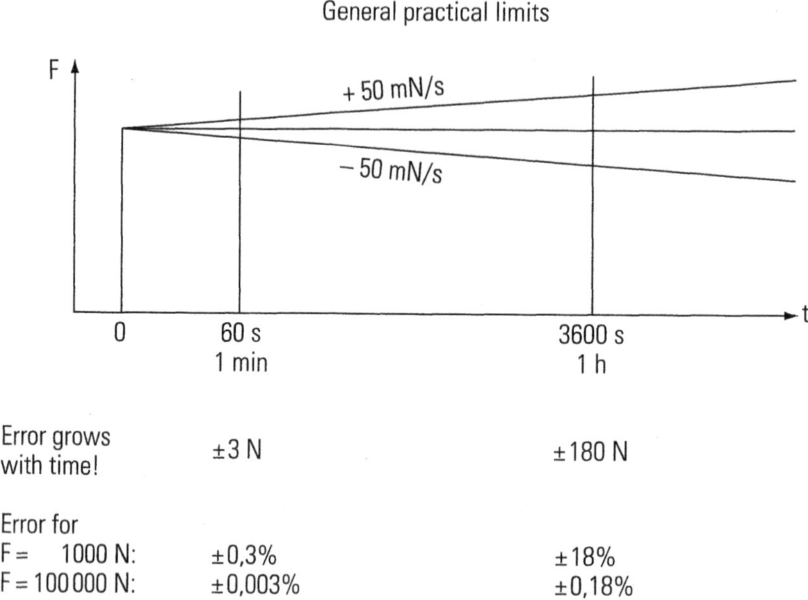

Figure 11.11 Practical meaning of the "quasistatic limit" in measuring force with a charge amplifier in DC-mode

In practice, as a rule of thumb, a drift of about 5 times the value as calculated above, i.e. about ±50 mN/s for a load washer, is a realistic and reliable value. Measuring a force of 1 kN or 100 kN over a time span of 1 minute or 1 hour – using the sensor like a balance – will be influenced by drift as shown in Fig. 11.11. After 1 minute, the error caused by drift will be within ±3 N, which corresponds to ±0,3 % for 1 kN and within ±0,003 % for 100 kN, i.e. in both cases, the error is within acceptable limits for most practical applications. After 1 hour, the error due to drift grows to ±180 N, which correspond to ±18 % for 1 kN (obviously unacceptable!) and to ±0,18 % for 100 kN, which still is fully acceptable.

The conclusion from this example is the general rule, that small measurands can be measured quasistatically – or in near DC mode – only over a short period of time (in the order of minutes) while large measurands can be measured quasistatically over longer periods of time (up to several hours).

11.6.5.2
Time Constant of the Feedback Circuit

The output voltage U_0 of the charge amplifier lies directly across the feedback (range) capacitor C_r. Because the insulation resistance of that capacitor is not infinitely high, a leakage current flows and discharges the capacitor exponentially. (see 10.2.2).

Table 11.3 Time constants of feedback capacitors

C_g in nF	0,01...1	2	5	10	20...
R_g in TΩ	>100	>50	>20	>10	>5...
τ in ks	1...100		all >100		

The time constant τ of a capacitor depends on the magnitude of its capacitance and the properties of the material constituting its dielectric. Range capacitors used in charge amplifiers usually have polystyrene as dielectric, which exhibits the characteristics summarized in Table 11.3. In capacitors smaller than about 1 nF the dielectric memory effect must be considered too, which may cause additional measuring errors.

11.6.5.3 *Dielectric Memory Effect in Capacitors and Cables*

In an electrically charged capacitor the electric charge will not remain only on the electrodes but a certain amount of free electrons (or "holes") will "diffuse" into the dielectric. This "diffusion" is very slow because of the high insulation resistance of the dielectric. Through this phenomena the capacitance of the capacitor seems to increase and the resulting voltage is lower. After the capacitor has been discharged, the phenomena reverses and the capacitor will continue to yield electric charge which is "diffusing" out of the dielectric again, until the free charges are balanced.

This apparent additional capacitance ΔC_r defines, together with the insulation resistance R_f, the time constant $\tau_{\Delta C}$ which describes the dielectric memory effect (Fig. 11.12). This time constant can be from a few seconds up to several minutes or even hours and therefore interfere with quasistatic measurements of small measurands. The apparent additional capacitance is more noticeable with small feedback capacitors, as shown in Table 11.4.

Manufacturers of capacitors usually specify much smaller values for the dielectric memory effect. The reason is that the standard test standards utilize circuits with far too small insulation resistance so that the real existing memory effect is not measured at all.

The dielectric memory effect can also be provoked when e.g. the insulation resistance of cables is checked. A coaxial cable is simply a capacitor, too, with the insulator (usually PTFE) as the dielectric. If an insulation tester (e.g. a teraohm meter) with a high test voltage (above about 100 V) is used, such cables can exhibit

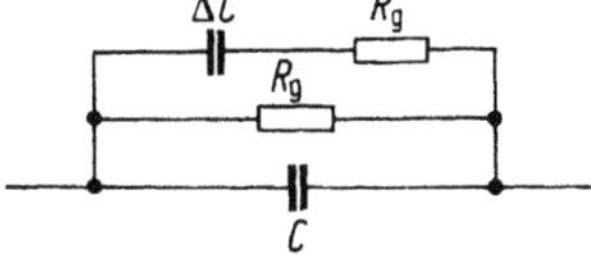

Figure 11.12 Substitute circuit for the dielectric memory effect

Table 11.4 Dielectric memory effect and its time constant

C_g	ΔC in % C_g	R_g in TΩ	$\tau\Delta C$ in s
100 nF	<0,5	1	<500
1 nF	<0,5	100	<500
100 pF	0,5...2	100	50...200
10 pF	5...10	100	50...100

a very strong dielectric memory effect. For a considerable time electric charge seems to "leak" out of the cable which may disturb the measurement. Short-circuiting the cable will not eliminate that electric charge at once because of the particular nature of the dielectric memory effect. Therefore care should be taken to use a test voltage as low as possible when checking the insulation resistance of cables and other components of the input circuit of a piezoelectric measuring chain, especially when small measurands are to be measured.

11.6.5.4
Leakage Current at the Input

The blocking layers in the semiconductors used as input stage in charges amplifiers are not perfect insulators. Therefore such devices always have albeit very small but finite leakage currents in the order of fA. In addition, similar leakage currents may flow over the insulation materials in the input circuit (e.g. PTFE-insulated soldering points). These leakage currents are highly temperature-dependent and increase exponentially with temperature (except in MOS-FET). Leakage currents through the blocking layers double for about every 8 °C temperature rise.

High-quality devices exhibit the values given in Table 11.5 for their leakage current. These values are obtained under optimum conditions and may be easily higher by a factor of 10 in practice.

The leakage currents cause a drift in the output voltage which is constant if the temperature remains constant. The total leakage current I_l flowing through the input means that an electric charge $Q = I_l \cdot t$, growing linearly with time, flows into the feedback (range) capacitor until the amplifier saturates – either at the positive or the negative limit of its output range.

Table 11.5 Leakage currents of various input devices for charge amplifiers

	leakage current in fA at	
Input stage with	20 °C	50 °C
Varactor diodes	< 3	< 30
MOS-FET	< 10	< 10
J-FET	<100	<1000

11.6.5.5
Zero Point Stability

In an ideal charge amplifier, input and output voltage would be exactly zero when the reset switch that short-circuits the feedback (range) capacitor is closed. Various influences such as temperature changes aging, changes in the blocking layers of the input device (e.g. after having been slight overloaded only by a higher voltage which did not fully damage the layer) will lead eventually to some zero shifts. Many amplifiers, especially those intended for quasistatic measuring, have a zero adjustment built-in which allows to bring the output voltage exactly to zero again when the amplifier is on "reset". Amplifiers for dynamic measuring only have an additional capacitive coupling fitted which eliminates any zero offset caused by a DC-component. The effect on the measuring result is discussed in 11.5.5.5.

The magnitude of such variations is about ±5 mV over several days for amplifiers with a MOS-FET differential input. If suddenly a larger deviation appears at the input, possibly combined with a higher leakage current, it is most likely that the MOS-FET has been "partially damaged". Such "partial damage" can result from a slight over voltage at the input, too small to fully damage the MOS-FET. To avoid such "partial damage" which, by the way, may go unnoticed for quite a time, it is advisable to carefully follow the safety and operating recommendations of the manufacturers, such as short-circuiting cables before connecting them to the amplifier, and so on.

Charge amplifiers with J-FET and varactor diodes at their input exhibit much smaller changes in the zero, i.e. only about ±1 mV over several days. Also, these devices are much less prone to "partial damage" than MOS-FETs. However, J-FETs have higher leakage currents and varactor diodes have a much lower upper frequency limit.

11.6.5.6
Leakage Currents Over the Insulation Resistance in the Input Circuit due to Offset Voltages

The input voltage of an ideal charge amplifier would be exactly zero, i.e. no voltage would be found in the cable and the sensor. This would mean that even with poor insulation resistance in the sensor and cable, no leakage current could flow. In the real charge amplifier, however, there is always a small voltage offset present at the input, which allows a small leakage current to flow. Such an offset voltage causes a leakage current to flow, which of course increases with lower insulation resistance in the input circuit. This offset voltage can be positive or negative and is within about ±20 mV for MOS-FET differential input stages. The resulting leakage current lets the measuring signal drift away. If the insulation in the input circuit is e.g. only 100 GΩ – a rather poor value – and the offset voltage is e.g. –10 mV, a drift current of $I_d = -10\,\text{mV}/100\,\text{G}\Omega = -100\,\text{fA}$ results. This corresponds to a charge drift of –100 fC/s, applying the relationship $I = Q \cdot t$.

In order to measure with a minimal drift it is important to minimize the offset voltage at the amplifier input. In charge amplifiers with built-in zero adjustment, this can be done as follows: First, the complete system is set up for the particular measurement to be made, including setting the needed measuring range, and the shortest time constant available is selected. Then, the input is short-circuited with lets the amplifier saturate immediately. Carefully adjusting the zero will lead to a point where the amplifier flips from positive into negative saturation or vice versa. This "flipping" point defines the best possible adjustment for minimal drift in the given measuring setup. Obviously, the amplifier must first be set to DC-mode (quasistatic setting, "long time constant") again before starting the measurement.

Although the state so adjusted will not remain for a very long time because of the limited stability of the zero, it nevertheless makes quasistatic measuring over quite long periods of time possible, even with rather poor insulation resistance. This procedure is – of course – only realistic for making special measurements in the laboratory where the limit must be pushed as far as possible.

11.6.5.7 *Output Voltage with Poor Insulation Resistance at the Input and Short time Constant*

Contrary to the example described in 11.5.5.5, using short time constant means that a resistor R_f smaller than the insulation resistance by several orders of magnitude is switched parallel with the feedback (range) capacitor.

Again the offset voltage causes a leakage current to flow over the insulation resistance in the input circuit. This charges the capacitor C_f, but only until an equal but opposite current flows over the insulation resistance of the range capacitor, which ends in a state of equilibrium with the error voltage U_{of} at the output of

$$U_{af} = U_{offset}\left(1+\frac{R_g}{R_{tot}}\right). \tag{11.17}$$

The output voltage will approach asymptotically the value U_{of} after setting the amplifier to measuring ("operate"), following the time constant $\tau = R_f \cdot C_f$ (Fig. 11.13).

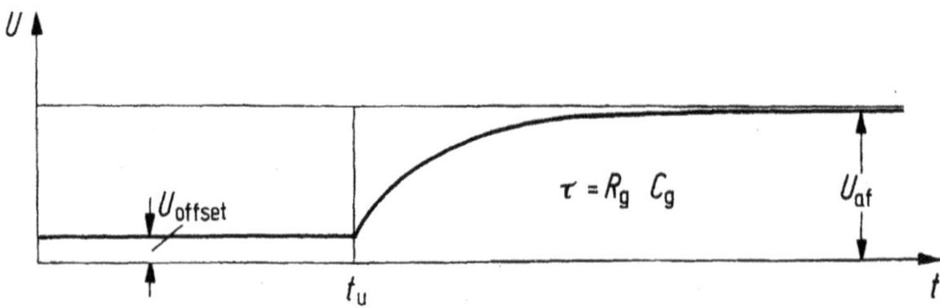

Figure 11.13 Output voltage with poor insulation resistance at the input and short time constant, before and after switching from "reset" to "operate" at the instant t_0

11.6.5.8
"Operate Jump"

In many charge amplifiers, especially those intended for quasistatic measuring, the feedback (range) capacitor can be "short-circuited", i.e. discharged, with a reset switch parallel to it. It must be remembered here that it is not a true short circuit (this would damage the operational amplifier!), but an extremely short time constant, defined by a resistor of about 10kΩ in series with the reset switch. Resetting defines the zero point for the measurement to be made immediately afterwards. Although the voltage across the feedback (range) resistor has dropped to zero during reset, when opening the reset switch to "operate", a small shift or "jump" is often observed in the output voltage.

This jump in output voltage, often called "operate jump" in practice, stems from two sources: the inherent noise of the operational amplifier and the residual electric charge in the reset switch. However, the "operate jump" is only of concern when making quasistatic measurement, i.e. with the amplifier set to DC-mode ("long time constant").

All amplifiers have some inherent noise, i.e. its zero fluctuates stochastically around an average value over a wide spectrum of frequencies. With high probability the instantaneous value of the zero point is not exactly zero when switching to "operate". Therefore the output voltage will not remain at zero but "jump" to a value corresponding to the instantaneous value of the noise voltage at the instant of switching.

This "jump" is different each time and its magnitude can not be predicted, except that it will remain within certain limits according to the peak values of the noise voltage. Therefore operational amplifiers with the lowest possible noise level are chosen for charge amplifiers.

The second cause for the "operate jump" can be the reset switch. This switch must be highly insulating, at least at the side which is connected to the amplifier input. Electric charge that was generated by operating the switch through friction (triboelectric effect) may stay for a long time on the highly insulating parts of the switch and create an electric field. If the switch is operated again, a charge shift may occur which is registered by the charge amplifier like a real measuring signal. Even

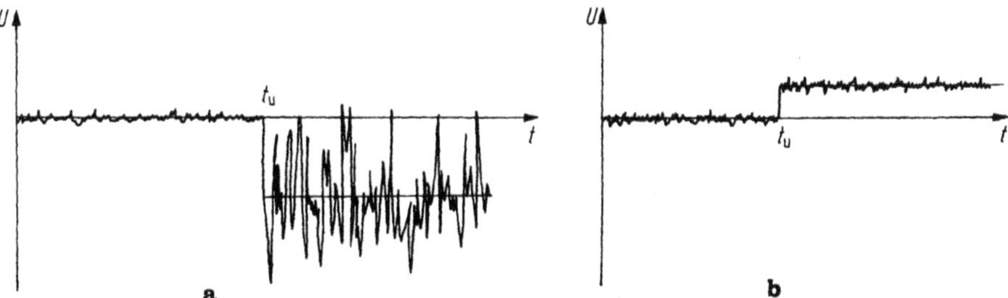

Figure 11.14 "Operate jumps", provoked by the input noise of the amplifier **a**, and by moving the switch contact in an electrostatic field or by residual electric charge on the reeds in the reed relay **b**. They occur at the instant t_0 when switching from "reset" to "operate"

when a reed relay is used as reset switch, therefor may be residual charge on the contact reeds in the relay.

Unlike the "operate jump" caused by noise, it mostly is very much the same each time and can be positive or negative. With suitable reset switches the "operate jump" can easily be kept below a value corresponding to 1 pC. Both types of "operate jump" are illustrated in Fig. 11.14.

When using certain charge amplifiers, especially older types which have a discrete time constant switch, the "operate jump" can be virtually eliminated by using the trick of not resetting the amplifier with the reset switch but by setting the time constant switch to "short", which in effect will also reset the amplifier, and then set it back to "long" before starting the measurement.

11.6.5.9
Equalization of Electric Charge After Manipulations

As shown already in 11.5.5.3, excess voltage at the input may partially damage a MOS-FET. This can happen if e.g. a cable is connected whose insulation was tested just shortly before and which is still electrically charged to a certain voltage. Also when measuring with a very small range capacitor and very short cables, voltage peaks at the input may occur in response to large measurements with steep rise times. In the latter case the operational amplifier may not be able to keep the input at zero if its slew rate is insufficient and the voltage may momentarily have a peak which goes into the critical range of about 25 to 60 V which can result in a "partial damage". After such a "partial damage"has occurred, a abnormally large zero shift will be noticed and drift may also increase considerably. Then it is advisable to change the MOS-FET.

11.6.6
Influence of Cables

In general the cable has no influence when measuring with charge amplifiers. However, if the cable capacitance C_c is very high (i.e. the cable is very long) and the inner gain v_i of the operational amplifier not sufficiently high, the amplitude of the signal will be reduced by the factor

$$k=\frac{1}{1+\frac{1}{v_i}\left(1+\frac{C_k}{C_g}\right)}. \tag{11.18}$$

e.g. a cable of 100 m length and a capacitance of 70 pF/m has a total capacitance of 7 nF. Using a small feedback (range) capacitor of e.g. 10 pF and an older amplifier with an inner gain of only vi≈1000, the reduction factor will work out as k=0,588, i.e. the signal amplitude would be nearly halved. With a modern amplifier having a gain $v_i \approx 100000$, the reduction factor would be only k=0,993, i.e. the signal reduced by less than 1 %. Moreover this is an extreme case with such a long cable and a small range capacitor.

Far more critical with long cables is the noise at the amplifier input. Current MOS-FET input stages have a noise of about 200 µV_{pp}. This noise will be amplified by about the factor C_c/C_r, which comes to C_c/C_r=700 in the example cited above. The noise level would therefore rise to about 140 mV_{pp} already.

With increasing length, the cable will also start to influence the upper frequency limit. This is described in 11.5.4.

11.6.7 Properties of Currently Used Input Stages

Devices used as the input stage of a charge amplifier must have an extremely high insulation resistance and an extremely small leakage current. Despite the progress in electronics, the mainly used devices remain the electrometer tube, varactor diode, MOS-type field effect transistor (MOS-FET) and junction-type field effect transistor (J-FET). Some other devices may be used sometimes but not for charge amplifiers used for measuring.

For measuring quasistatically the three devices mentioned first are most suitable because of their small leakage current. However, electrometer tubes are rarely used anymore because they are bulky, are sensitive to shock and show a pronounced microphone effect, i.e. they pick up noise and vibration, too.

Varactor diodes are best in terms of the required specifications. Their main drawback is their rather low upper frequency limit (because of the chopper principle used) and their relatively high cost.

MOS-FETs as input stage are still the preferred choice for charge amplifiers with a wide frequency range from quasistatic to far over 100 kHz, despite their limited stability of zero and proneness to damage through over voltage.

J-FETs are primarily used for dynamic measuring because of their bigger leakage current which limits the possibilities of quasistatic measuring. Moreover their leakage current increases rapidly with temperature (doubling for every 8 °C). A very good J-FET has a leakage current 10 times of that of a MOS-FET at 22 °C. At 50 °C, the leakage current of a J-FET is already 100 times bigger.

The characteristics of the most commonly used input devices are summarized in Table 11.6.

Table 11.6 Comparision of input stages with J-FET and MOS-FET

Application	J-FET	MOS-FET
Dynamic	good	good
Quasistatic	only for large ranges	good
Risk of damage from overvoltage	small	great
Zero stability	good	sufficient
Error after overload	yes	no
Leakage current at higher temperature	large	small
Sensitivity to radiation	low	high

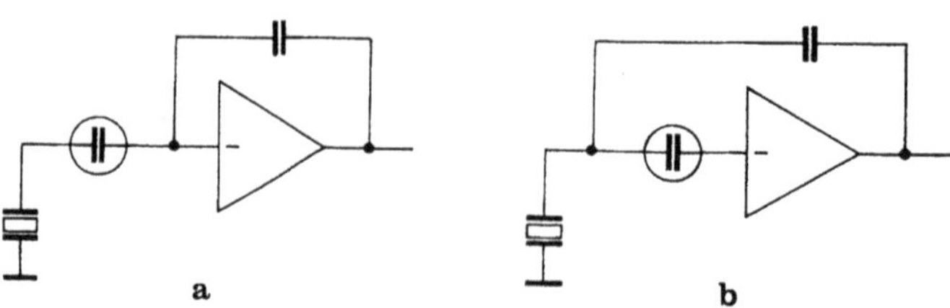

Figure 11.15 Capacitive coupling at the input **a** or within the feedback circuit **b** of a charge amplifier

11.6.8 Capacitive Coupling for Measuring at High Temperature

Piezoelectric sensors often generate an electric voltage at high temperature, even under stationary conditions and no change in measurand. This voltage can be a thermoelectric effect (usually less than about 10 mV) or of galvanic nature (up to about 500 mV). At the same time, the insulation resistance of sensor and cable tends to drop rapidly above 200 °C.

This voltage produces a leakage current across the reduced insulation resistance and a strong drift in the charge amplifier output, unless a resistor of sufficiently low value is switched in parallel with the feedback (range) capacitor. That in turn means a very short time constant and a correspondingly higher lower frequency limit.

The voltage mentioned above can be kept away from the charge amplifier by capacitive coupling with a capacitor inserted before the input of the amplifier (Fig. 11.15) which prevents a leakage current from flowing. However such capacitive coupling changes the frequency characteristics at the lower frequency limit considerably and each case must be carefully checked by actually measuring the frequency characteristics or by calculation to avoid measuring errors.

11.6.9 Protecting the MOS-FET at the Input from Overload

MOS-FET have the shortcoming that they are sensitive to overload by excessive voltage. At more than 100 V a MOS-FET is usually destroyed by a breakdown of the blocking layer. While a voltage less than about 50 V will cause no damage, in a range from about 50 to 100 V, the MOS-FET may suffer a so-called "partial damage", i.e. its blocking layer may be weakened which shows in a suddenly much larger zero offset and an increased leakage current.

In principle, the voltage at the input remains zero during normal operation. However, there are certain situations where a voltage which is critical may be develop. The most common cause for damage by an excessive voltage is connecting cables which are still electrically charged. Cables that have been moved vigorously can easily be electrically charged to over 100 V through triboelectric effects between conductor and insulating material. Also it is often overlooked that some insulation testers apply a measuring voltage of more than 50V. A cable whose insulation was checked by such an instrument will stay electrostatically charged for quite a while – discharge time constants may exceed several hours – and obviously damage the MOS-FET when being connected to the charge amplifier input. Also a sensor can sometimes be the cause when a large measurand is applied shortly before the sensor is connected. The open output voltage of a sensor, especially of force sensors, can easily reach several hundred Volt.

So far no satisfactory method has been found to protect the MOS-FET against overload without severe drawbacks. Moreover it should be remembered that – in contrast with electrometer amplifiers – even a closed reset switch does not entirely protect the MOS-FET. Therefore the best protection is obtained by consistently following the practice of first connecting the sensor to the cable and then always short-circuiting the cable just before connecting it to the charge amplifier.

Damage through over voltage can also occur during operation under certain circumstances, especially when the charge amplifier is set to a small measuring range (small feedback capacitor) and the connected sensor yields an extremely high electric charge (e.g. in response to a large measurand value). The output voltage of the amplifier is limited and the amplifier can not – with the small range capacitor – generate a sufficiently large electric charge to compensate the charge yielded by the sensor. The voltage at the input will not remain zero and may briefly reach critical values.

Such a problem may also occur during certain vibration measurements. If vibration at low frequencies and of small amplitude are to be measured and at the same time the natural frequency of the acceleration sensor is excited by some high-frequency phenomena and then the sensor is yielding correspondingly large electric charge. In such situations the problem can often be solved by inserting a low-pass filter at the input of the amplifier to filter out such large electric charge coming from the natural frequency of the sensor yet allowing the low-frequency signals of small amplitude to be measured correctly.

Although the reader may by now be quite apprehensive about damaging a MOS-FET when using a charge amplifier, the author can state that in over 25 years of experience and working in the field, he has never "succeeded" in damaging a MOS-FET – despite ignoring the recommended precautions most of the times. Therefore, as practice shows, with reasonable care few problems will ever be encountered in the field.

11.7 Connecting Several Sensors to One Amplifier

Piezoelectric sensors are active sensors (see 11.2.1) and therefore, several sensors can be connected in parallel to a single amplifier. Thereby it is essential that all sensors so connected have the same sensitivity. Usually, sensors that are connected as a pair or a group to a single amplifier are of the exactly the same type, i.e. they are selected from a batch to have sensitivities within very close tolerances. It must be remembered that although the sensitivity of a piezoelectric sensor – especially a quartz sensor – is essential determined by the piezoelectric coefficient of the piezoelectric transduction element (and therefore is always the same for a given model and design of a sensor), there is a certain variation in sensitivity because the sensor housing, which usually forms a force shunt, has also certain tolerances in dimension (machining tolerances) and its elasticity may also slightly vary due du variations in material properties of the housing. If the sensors are preloaded in the application, then they must be selected to have the same sensitivities in the preloaded state (i.e. the calibration values obtained from sensors preloaded in a standard calibration rig).

With sensors of selected and thus quite equal sensitivity, the output obtained from the amplifier is the algebraic sum of all measurands acting on all connected sensors. If an electrometer amplifier is used, the complete measuring system (including cabling) must be calibrated in order to determine the valid sensitivity of the system. If cables are exchanged for cables of different lengths, calibration must be repeated (see 11.3).

With a charge amplifier, it is sufficient that all connected sensors have the same sensitivity while their respective capacitance and the capacitance of the respective cables (cable length) may quite differ without affecting the measuring result (see 11.4).

The method of connecting several sensors in parallel to a single amplifier is mainly used with dynamometers and force plates (see chapter 6), but can be used effectively in any situation where only the sum off all measurands acting on a number of sensors is of interest. This possibility not only allows economic solutions but also makes operation of such systems very easy because only one amplifier must be controlled although any number of sensors may be connected (see e.g. 6.2.2, 6.3 and 6.7).

11.8 Cables and Connectors

Cables for connecting piezoelectric sensors to charge amplifiers and electrometer amplifiers must meet several requirements in order to obtain good measuring results. They must have a sufficiently high insulation resistance, be as free as possible from triboelectric effects, have a small capacitance and maintain those characteristics over a sufficiently wide temperature range. Best suited and most commonly used are coaxial cables.

For quasistatic measuring an insulation resistance of over 100TΩ is required which is easily obtained in cables having PTFE as insulation material. Even after prolonged use, and with proper care, values of over 10TΩ can be maintained in the field. This requires that the connectors are kept scrupulously clean which is achieved by systematically covering them with appropriate protective covers (usually supplied by the manufacturer) when not connected to a sensor or an amplifier. The same applies to the connectors on the sensors and amplifiers.

Connectors are best cleaned with white gasoline or pure benzene. Some manufacturers also supply suitable special cleaning sprays, however the ordinary "electronic cleaning sprays" found in the general commerce must never be used because they are totally unsuitable and will completely deteriorate the high insulation. The FCW (Freon) sprays popular in the past must not be used any more for obvious reasons. A clean brush or non-fluffing paper tissue may be used to rub off dirt after spraying it with the cleaning agent. Compressed air as commonly used in industry is not suitable because it usually is not oil-free and will deteriorate the insulation. Also blowing with the mouth must be avoided because of the saliva droplets. Best for blowing dirt out of connectors is an ear syringe made of rubber.

Whenever possible cables should be laid out and fixed in such a way that they will not move during measuring. In many applications the sensor is mounted on a moving part and the cable will be moving around or bending and stretching, especially when measuring shock and vibration. Therefore cables should be free of triboelectric effects, i.e. they should not produce electric charge when moved or bent. Ordinary coaxial cables, even with PTFE insulation, will easily produce an electric charge when moved – an amount of charge that can easily exceed that yielded by the sensor! Therefore only so-called "low-noise cables" should be used. Such coaxial cables have a graphite layer between the PTFE insulator and the wire-braided screen, preventing an electric charge from being generated by the triboelectric effect when the screen rubs against the PTFE. Formerly oil was also used instead of graphite which was very effective in preventing the triboelectric effect. However, the oil tended to dry out after some time, or often it simply leaked out – soiling the insulating part of the connectror on the way – and the protection was lost. Therefore oil-filled cables are no longer in use.

Thin coaxial cables are generally less prone to produce noise when moved and should be preferred when measuring on moving objects. At the same time their mass and cable capacitance are lower.

There is a large number of coaxial connectors that are used in piezoelectric systems which are described in 5.5.5. Some of these connectors, especially those that are not closed by screwing both sides together (e.g. BNC connectors) may also produce unwanted electric charge when moved or when being plugged together. Therefore all connections should be made before setting the amplifier to operate and connectors should be fixed so they will not move during measuring. This is particularly important when measuring small measurands. Screw-on connectors are preferable but it must be assured that they will not become loose e.g. through vibration during measuring. Safety-wiring or applying a shrink sleeve are effective measures to prevent this.

11.9 Calibration of Charge Amplifiers

The error in converting electric charge into electric voltage by a charge amplifier is primarily determined by the uncertainty of the value of the feedback (range) capacitor. Properly selected and aged capacitors generally have a capacitance with an uncertainty of within ±1% for values over 50pF and up to ±3% for smaller values. Of course errors in gain of the voltage amplifier stage usually following the charge amplifier stage proper and the linearity of that stage must be considered, too. However with proper design, these errors can easily be kept at least one order of magnitude smaller than the uncertainty – and stability – of the range capacitors. A charge amplifier can only be calibrated by applying a precisely known electric charge to its input and measuring the corresponding output voltage. For generating such a "calibration charge" a precise voltage source is connected to the charge amplifier input via a precisely known and highly insulating capacitor. Or the calibration charge is obtained by loading a precisely calibrated sensor with a precisely know measurand. This latter method is also known as "in situ calibration" and is usually performed in a completely installed and assembled measuring chain. Often the connected indicators and recorders are directly included in this type of calibration,also called "over all" calibration. As far as the charge amplifier is concerned, this type of calibration is identical as the now described electric calibration – in both cases, a known electric charge is applied to the input – and the same points have to be observed.

Electric charge can be generated by switching a voltage source in series with a capacitor. According to the relationship $Q=U\cdot C$, the error of the electric charge Q obtained is the combination of the errors in the voltage U and in the capacitance C measurement. The error of the voltage measurement can easily be kept orders of magnitudes smaller than the capacitance measurement. Fig. 11.16 shows the usual circuit which allows to switch the capacitor either to ground or to the voltage source. The capacitor C must be highly insulating because its insulation resistance R_{is} is between the ground or the low output resistance of the voltage source and the charge amplifier input. After switching to the voltage source, adjusted to the required voltage U, either the charge amplifier output voltage can just be recorded (this is a "calibration") or the amplifier can be set to obtain the desired conversion factor between applied electric charge and output voltage by adjusting its gain (this is an "adjustment").

Although this kind of calibrating appears to be a simple procedure, there are several effects that have to be considered very carefully in order to avoid errors For calibrating quasistatically, a DC constant voltage is used. When switching from ground to that voltage (Fig. 11.17) an ideal step in electric charge is obtained. The electric charge so generated remains as long the voltage source is switched to the capacitor.

When measuring the output voltage of the charge amplifier immediately before the step in the electric charge at the instant t_1 and immediately after it at t_2, the

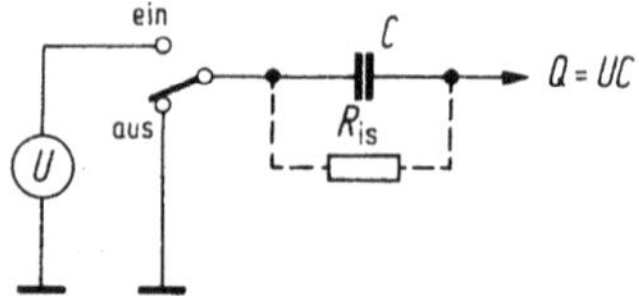

Figure 11.16 Basic circuit for producing a step-like change of electric charge

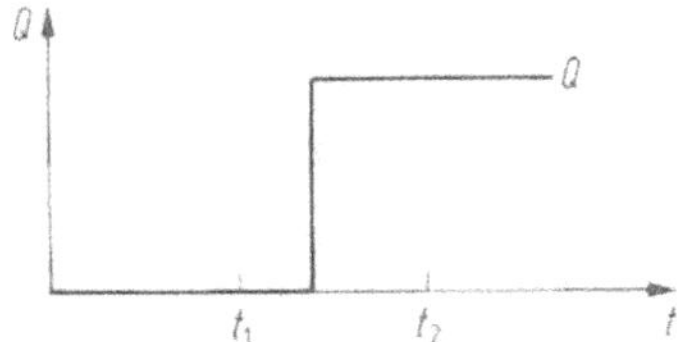

Figure 11.17 Ideal step-like change in electric charge

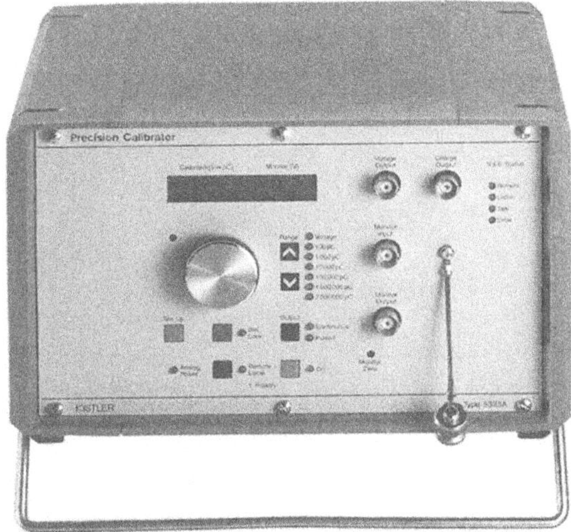

Figure 11.18 Precision charge calibrator (Courtesy of Kistler)

measured change in output voltage corresponds to the electric charge applied. During the finite time interval t_2-t_1, the inherent drift of the charge amplifier (see 11.5.5) has produced a change in output voltage, leading to an error of the calibration or adjustment. The error stemming from drift grows linearly with time and can therefore be minimized by proceeding quickly. The inherent drift of the charge amplifier is increased by the leakage current $I=U/R_{is}$ flowing over the insulation resistance of the calibration capacitor. Under stable conditions it is quite feasible to measure the drift and compensate for its influence by also measuring the time interval t_2-t_1.

If the rise time of the step in electric charge is extremely short (e.g. when just a switch is used), the upper frequency limit of the amplifier to be calibrated may be not high enough to reproduce this step signal without overshoot or even brief saturation. Therefore when setting up a calibration system, the signal should be monitored with an oscilloscope to make sure that such phenomena will not result in calibration errors. Often the rise time of a calibrator must be chosen long enough to avoid such problems, but this tends to lengthen the time interval t_2-t_1

again. It becomes clear that calibrating quasistatically becomes difficult when insulation resistance is low and when small feedback (range) capacitors in the charge amplifiers are used. This is so especially in charge amplifiers used for dynamic measuring only such as in measuring vibrations.

These problems can be avoided by using an AC instead of a DC voltage. Then the frequency should be chosen to lie in that part of the frequency range of the charge amplifier to be calibrated that is not influenced in amplitude and phase response by neither the lower nor the upper frequency limit nor any filters.

Several manufacturers offer specially designed charge calibrators which usually have all the necessary features to avoid the problems mentioned above. The precision charge calibrator shown in Fig. 11.18 generates calibrations signals of optimized shape and can be combined into PC-controlled calibration systems.

References

Aitchison, Leslie: A History of Metals. Vol. 1, New York: Interscience Publishers, Inc. (1960), p 78

Aizu, K.: Phys. Rev. B 2 (1970) 754

Aizu, K.: J. Phys. Soc. Japan 32 (1972), 1287 and 34 (1973) 121

Arnold, H.: Displazive Phasenumwandlung, Gitterdynamik und Röntgenbeugung. Diss. Fak. f. Bergbau und Hüttenwesen. TH Aachen 1976

Aschero, G., Gizdulich, P., Mango, F.; Romano, S.M.: Converse Piezoelectric Effect Detected in Fresh Cow Femur Bone. J. Biomechanics, Vol. 29, No. 9, 1169-1174, 1996

ASTM E1106-86: Standard Method for Primary Calibration of Acoustic Emission Sensors. (1986). Annual book of ASTM standards, Vol. 03.03, ASTM , Philadelphia, PA. See also: Method E1106-86(1997): Standard Method for Primary Calibration of Acoustic Emission Sensors; Guide E650-97 Standard Guide for Mounting Piezoelectric Acoustic Emission Sensors); and: Guide E976-00 Standard Guide for Determining the Reproducibility of Acoustic Emission Sensor Response

Ballman, A.A.: J. Crystal Growth 1 (1967) 37

Bao, Min-Hang: Micro Mechanical Transducers Pressure Sensors, Accelerometers and Gyroscopes. Amsterdam: Elsevier 2000

Barret, R.M.: Very Low Frequency Accelerometer Measurements with Piezoceramic Sensors. Proceedings SENSOR '95, Kongressband p 183-187

Bauer, A.; Bühling, D.; Gesemann, H.J.; Helke, G.; Schreckenbach, W.: Technologie und Anwendungen von Ferroelektrika. Leipzig: Akad. Verlagsges. 1976

Baumgartner, H.U.; Gautschi, G.H.: Piezoelectric Strain Transducers. London: Transducer '80 Conference 1980

Bechmann, R.: Phys. Rev. 110 (1953) 1060

Bechmann, R.; Ballato, A.D.; Likaszek, T.J.: Proc. IRE 50 (1962) 1812

Bechmann, R.; Hearmon, R.F.S.: Elastische, piezoelektrische, piezooptische und elektrooptische Konstanten von Kristallen. Landolt-Börnstein, Neue Serie (Hrsg. K. H. Hellwege), Bd. III/1. Berlin, Heidelberg, New York: Springer 1966

Bechmann, R.; Hearmon, R.F.S.; Kurtz, S.K.: Elastische, piezoelektrische, piezooptische elektrooptische Konstanten und nichtlineare dielektrische Suszeptilitäten von Kristallen. Landolt-Börnstein, Neue Serie (Hrsg. K. H. Hellwege), Bd. III/2. Berlin, Heidelberg, New York: Springer 1969

Berlincourt, D.A.: Piezoelectric Crystals and Ceramics. Ultrasonic Transducer Materials (Publ. O. E. Mattiat). New York: Plenum Press 1971

Berlincourt, D.A.; Curran, D.R.; Jaffe, H.: Piezoelectric and Piezomagnetic Materials. Physical Acoustics (Publ. W. P. Mason, vol IA. New York, London: Academic Press 1964

Bertagnolli, E.: Untersuchung der sekundären Zwillingsbildung in α-Quarz. Diss. Naturwiss. Fak. Univ. Innsbruck 1979

Bertagnolli, E.; Kittinger, E.; Tichý, J.: Sekundäre Zwillingsbildung in X-Quarzplatten. Jahrerstagung Oesterr. Phys. Ges. Loeben 1977

Bertagnolli, E.; Kittinger, E.; Tichý, J.: Le Journal de Physique – Lettres 39 (1978) 295

Berther, T.; Kubler, J.; Gautschi, G.H.: K-Shear Accelerometers. Noise & Vibration Worldwide, Oxford Elsevier Advanced Technology, June 1997

Berther, T.; Burhard, H.: PiezoelectricMeasuring Systems for Vehicle Development. Kistler company publication 1999 (available as reprint 20.203e from Kistler)

Berther, T.; Jacobson, M.: New Possibilities to Measure Force at Crash Test Barriers. Kistler company publication 2000 (available as reprint 20.210e from Kistler)

Bezdelkin, V.V.; Antonova, E.E.: Proc. 1st European Workshop on Piezoelectric Materials, Montpellier, France, 1983, Journal de Physique IV suppl. C2, Vol. 4 (1994) C2-139

Bhalla, A.S.; Cross, L.E.; Whatmore, R.W.: Proc. 6th Int. Meeting on Ferroelectrics, Kobe, Japan, 1985, Jap. J. Appl. Phys. Supplement 24-2 (1985) 727

Bill, B.: PiezoBEAM – ein neuartiges Konzept für Beschleunigungssensoren (PiezoBEAM-Accelerometers: a Proven and Reliable Design for Modal Analysis). Messen & Prüfen 26 (1990) vol 3 (available as reprint 20.148e from Kistler)

Bill, B.: Basic Theory of the Hammer Test Method. Kistler company reprint 20.195e, 1998

Bill, B.; Wicks, A.L.: Measuring Simultaneously Translational and Angular Acceleration with the New Translational-Angular Piezobeam (TAP) System. Elsevier Sequoia: Sensors and Actuators A 21-A23 (1990) 282-284

Bill, B.; Engeler, P; Gossweiler, C.: Inertia Compensated Force and Pressure sensors. Proc. SENSOR 01 Conference, Nürnberg, Germany, 8-10 May 2001, paper A4.5

Bohaty, L.: Quadratische elektrostriktive Effekte in Kristallen. Diss. Math.-Naturwiss. Fak. Univ. Köln 1975

Brice, J.C.: Quartz for Resonators. Rev. Mod. Phys. 57 (1985) 105-146

Bridgman, P.W.: J. Appl. Physics 12 (1941) 461

Briot, H.E.; Bigler, E.; Daniau, W.. Marianneau, G.; Pafkar, A.: A Comprehensive Mapping of Surface Acoustic Wave Properties on Gallium Orthophosphate ($GaPO_4$). Proc. 13. Eur. Frequency and Time Forum / Int. Frequency Control Symp., Besançon: 12-16 April 1999

Broch, J.T.: Die Anwendungen der B & K Messsysteme für Messungen von mechanischen Schwingungen und Stössen. Nærum: Brüel & Kjær 1970

Brüel & Kjær: Piezoelectric Accelerometers and Vibration Preamplifiers (company publication). Nærum: Brüel & Kjær 1976

Brüning, D.M. von: Experimenteller Nachweis des elektrokalorischen Effektes in Turmalin zwischen 9 K und 24 K. Diss. Phil. Fak. II Univ. Zürich 1969

Busch-Vishniac, I.J.: Electromechanical Sensors and Actuators. New York: Springer 1999

Burkard, H.; Calame, C.: Rotating Wheel Dynamometer with High Frequency Response. Tire Technology International, 1998, p 154 (available as reprint 20.192e from Kistler)

Burkard, H.; Berther, T.: New Rotating Wheel Torquemeter (RTW) for Driving and Braking Torque Measurements. Kistler company publication 2000. (available as reprint 20.208e from Kistler)

Cady, W.G.: Phys. Rev. 17 (1921) 531; 19 (1922) 1; Proc. IRE 10 (1922) 83

Cady, W.G.: Piezoelectricity (An Introduction to the Theory and Application of Electromechanical Phenomena in Crystals). New York: Dover 1964

Calderara, R.: Swiss patent file 536561 (Kistler patent), Bern 1971

Calderara, R.: Long-Term Stable Quartz WIM Sensors. NATDAC '96 National Data Acquisition Conference, 9-11 May 1996, Albuquerque, NM, USA (available as reprint 20.178e from Kistler)

Calderara, R.; Baumgartner, H.; Tichý, J.; Sonderegger, H.: Swiss patent file 536489 (Kistler Patent), Bern 1971

Carl, K.: Eine Untersuchung von Polarisation, Domänenausrichtungsgrad und des Maximums der piezoelektrischen Aktivität an der morphotropen Phasengrenze von keramischen $PbTiO_3$-$PbZrO_3$-Mischkristallen. Diss. Fak. Elektrotechn. Univ. Karlsruhe 1972

Cavalloni, C.; Kirchheim, A.: New Acoustic Emission Sensors for In-Process Monitoring. Proc. 12th Intl. Acoustic Emission Symposium, Sapporo (Japan). Published by: The Japanese Society for Non-Destructive Inspection, 1994

Chai, B.; Lefaucheur, J.L.; Ji, Y.Y.; Qiu, H.: Proc. IEEE Int. Frequency Control Symposium, 1998, p 748

Cook, R.K.; Weissler, P.G.: Phys. Rev. 80 (1950) 712

COST: Weigh-in-Motion of Road Vehicles Final report). Appendix 1: European WIM specification, Version 3.0. Laboratoire Central des Ponts et Chaussées LCPC, Paris: 1999

Curie, J.; Curie, P.: C. R. Acad. Sci. 91 (1880) 294; Bull. soc. min. de France 3 (1880) 90

Détaint, J.; Capelle, B.; Zarka, A.; Palmier, D.; E. Philippot, E.: Frequency Control and Filtering Applications of the New Piezoelectric Materials of Class 32. Ann. Chim. Sci. Mat. 22, pp 651-660 (1997)

Dietrich, R.V.: The Tourmaline Group. New York: Van Nostrand Reinhold Company Inc. 1985: pp 1 and 166

Doebelin, Ernst O.: Measurement Systems. New York: McGraw-Hill 1989

Dolino, G.; Bachheimer, J.P.; Vallade, M.: Appl. Phys. Lett. 22 (1973) 623

Donnay, G.; Barton, R.: Absolute Orientation of the Tourmaline Crystal Structure. Carnegie Inst. Washington Yearbook 65 (1967) 299

Donnay, G.; Buerger, M.J.: Acta Crystallogr. 3 (1950) 379

Drouillard, T.F.: Anecdotal History of Acoustic Emission from Wood. J. of Acoustic Emission, vol 9, no 3 (1990), pp 153-176

Drouillard, T.F.: Acoustic Emission – the First Half Century. Proc. 12th Intl. Acoustic Emission Symposium, Sapporo (Japan). Published by: The Japanese Society for Non-Destructive Inspection, 1994

El Messiery, M.A.; Hastings, G.W. and Rakowski, S.: Ferro-Electricity of Dry Cortical Bone. J. Biomed. Eng. 1979, Vol. 1, January

Engeler, P.: Acceleration-Compensated Pressure Transducer. US Pat. No. 6 105 434, 2000

Epprecht, W.: Schweiz. Min. Petr. Mitt. 33 (1953) 481

Erhart, J.; Fousek, J.; Janovec, V.; Prívratská, J. and Tichý, J.: Fundamentals of Piezoelectric Sensorics. Berlin, Heidelberg, New York: Springer 2002

Ewins,D.J.: Modal Testing Theory, Practice and Application. Baldock: Research Studies Press 2000

Feldtkeller, E.: Dielektrische und magnetische Materialeigenschaften. Mannheim: Biblogr. Inst. 1974

Fischer, H.; Matthias, E.: Piezoelektrischer 3-Komponenten Schnittkraftmesser für statische und dynamische Messungen. Ber. Werkzeugmasch.-Lab. ETH Zürich, für CIRP, 1968

Förster, F.; Scheil, E.: Akustische Untersuchung der Bildung von Martensitnadeln (Acoustic Study of the Formation of Martensite Needles). Zeitschrift für Metallkunde, vol 29, no. 9 (September 1936), pp 245-247 (in German)

Fraden, J.: Handbook of Modern Sensors: Physics, Designs, and Applications. Woodbury, New York: AIP Press 1997

Frank, R.: Understanding Smart Sensors. Boston: Artec House 1996

Fritze, H.; Tuller, H.L.; Borchardt, G.; Fukuda, T.: Proc. Materials Research Society Fall Meeting, Materials for Smart Systems III, Ed. M. Wun-Fogle, K. Uchino, Y. Ito, and R. Gotthardt, Vol. 604 (1999)

Fukuda, E.; Yasuda, I.: On the Piezoelectric Effect of Bone. J. Phsys. Soc. Japan 12 (1957) 1158-1162

Fukuda, T.; Takeda, H.; Shimamura, K.; Kawanaka, H.; Kumatoriya, M.; Murakami, S.; Sato, J.; Sato, M.: Proc. 11th IEEE Int. Symposium on Applications of Ferroelectrics, Montreux, Switzerland, 1998, p 315

Galassi, C.; Dinescu, M.; Uchino, K.; Sayer, M. (eds): Piezoelectric Materials Advances in Science, Technology and Applications. Dordrecht: Kluwer Academic Publishers 2000

Ganschow, S.; Cavalloni, C.; Reiche, P.; Uecker, R.: Proc. 11th Conf. on Solid and Liquid Crystals: Materials Science and Applications, Zakopane, Poland, 1994, Proc. SPIE Vol. 2373 (1995) 55

Gassmann, H.R.: Reib- und Verschleissverhalten der Bremsbeläge von Tragseilbremsen (Friction and Wear Behavior of Brake Pads in Track Cable Brakes). Zurich: ETHZ, Institut für Bau- und Transportmaschinen 1979

Gautschi, G.H.: Cutting Forces in Machining and their Routine Measurement with Piezoelectric 3-component Dynamometers. Proc. 12th Int. Machine Tool Design and Research Conf., Manchester 1971. London: Macmillan Press 1972

Gautschi, G.H.: Technica 21 (1972) p 1871

Gautschi, G.H.: Piezoelectric Multicomponent Force Transducers and Measuring Systems. London: Transducer '78 Conf. 1978

Geber, the Works of. Englished by Richard Russell, 1678, A New Edition with Introduction by E. J. Holmyard, MA., D., Litt. London, England: J. M. Dent and Sons Ltd. and New York, NY: E. P. Dutton and Co. (1928): pp 50, 66, 69, 139 and 149

Gmelins Handbuch der anorganischen Chemie (Hrsg. Gmelin-Institut. Silicium Teil B, System Nummer 15, 8. Auflage. Weinheim: Verlag Chemie 1959

Gohlke, W.: Mechanisch-Elektrische Messstechnik. München: Hanser 1955

Gohlke, W.: Einführung in die piezoelektrische Messstechnik. 2. Auflage. Leipzig: Akad. Verlagsges. 1959

Gossweiler, C.; Cavalloni, C.: High Sensitivity and Dynamic Ratio: Sensors with a New Piezoelectric Material. Proc. SENSOR 99 Conference, Nürnberg, Germany, 18-20 May 1999, vol 2, p 29

Grave, H.F.: Elektrische Messung nichtelektrischer Grössen. Leipzig: Akad. Verlagsges. 1962

Grouzinenko, V.B.; Bezdelkin, V.V.: Proc. IEEE Int. Frequency Control Symposium, 1992, p 707

Hankel, W.G.: Abh. Sächs. 12 (1881) 457; Ber. Sächs. 33 (1881) 53

Harris, C.M. (ed in chief): Shock and Vibration Handbook . New York : McGraw-Hill 1996

Hayakawa, R.; Wada, Y.: Piezoelectricity and Related Properties of Polymer Films. Adv. Polym. Sci. 11 (1973) 1

Heising, R.A.: Quartz Crystals for Electrical Circuits. Toronto, New York, London: Van Nostrand 1946

Hornsteiner, J.; Born, E.; Fischerauer, G.; Riha, E.: Proc. IEEE Int. Frequency Control Symposium, 1998, p 615

Hruska, K.: IEE Trans. on Sonics and Ultrasonics SU-18 (1971) 1; SU-24 (1977) 54

Hruska, K.; Kazda, V.: Czech. J. Phys. B 18 (1968) 500

Hruska, K.; Khogali, A.: Czech. J. Phys. B 19 (1969) 1092; IEEE Trans. on Sonics and Ultrasonics SU-18 (1971) 171

Hull, G.F.: Some Applications of Physics to Ordnance Problems. Jour. Franklin Inst., vol 192, pp 327-347, 1921

Hurle, D.T.J.; Cockayne, B.: Handbook of Crystal Growth, Vol. 2a, Ed. D.T.J. Hurle. Amsterdam: Elsevier 1993, Chapt. 3

Hwang, T.W.; Whitenton, Eric P.; Hsu, Nelson N.; Blessing, Gerald V.: Acoustic Emission Monitoring of High Speed Grinding of Silicon Nitride, Ultrasonic, 38, 554-559, (1999)

IEEE Standard on Piezoelectricity (ANSI/IEEE Std 176-1987)

Ikeda, T.: Fundamentals of Piezoelectricity. Oxford: Oxford University Press 1996

Inoue, N.; Iida, A.; Kohra, K.: J. Phsys. Soc. Japan 37 (1974) 742

IRE Standard on Piezoelectric Crystals. Proc. IRE-37 (1949) 1378; 46 (1958) 764

IRE Standard on Piezoelectric Crystals. Proc. IRE-49 (1961) 1161

ISA-S37.1-1975 – (Reaffirmed 1982) – Electrical Transducer Nomenclature and Terminology (Formerly ANSI MC6.1-1975) . Research Triangle Park, NC 27709: Instrument Soc. Amer. 1982

ISA-RP37.2-1982 – (Reaffirmed 1995) – Guide for Specifications and Tests for Piezoelectric Acceleration Transducers for Aerospace Testing. Research Triangle Park, NC 27709: Instrument Soc. Amer. 1995

ISA-S37.10-1982 – (Reaffirmed 1995) – Specifications and Tests for Piezoelectric Pressure and Sound-Pressure Transducers. Research Triangle Park, NC 27709: Instrument Soc. Amer. 1995

ISO; BIPM; IEC; OIML: International Vocabulary of Basic and General Terms in Metrology, 1984

ISO et al: International Vocabulary of Basic and General Terms in Metrology. 1993, ISO, Geneva

ISO Handbook: Mechanical Vibration and Shock (2 volumes). Ed. 2, 1995, ISO, Geneva

ISO Standards Handbook: Quantities and Units. Ed. 3, 1993, ISO, Geneva

ISO 12713: Non-Destructive Testing – Acoustic Emission Inspection – Primary Calibration of Transducers. 1998, ISO, Geneva

ISO 12714: Non-destructive testing – Acoustic Emission Inspection – Secondary Calibration of Acoustic Emission Sensors. 1999, ISO, Geneva

ISO/IEC 17025:1999 General Requirements for the Competence of Testing and Calibration Laboratories. 1999, ISO, Geneva

Jaffe, B.; Cook, W.R. Jr.; Jaffe, H.: Piezoelectric Ceramics. London: Academic Press 1971

Jäger, J.; Schoppa, G.; Schultz, W.: The Standard Instruments for the PTB for the 1 GPa Range of Pressure Measurement. PTB-W-66. Braunschweig: PTB 1996

Joshi, M.S.; Kotru, P.N.: Jap. J. Appl. Phys. 7 (1968) 700

Joshi, M.S.; Vag, A.S.: Soviet Physics Crystallography 12 (1968) 573

Jost, K.-H.: Röntgenbeugung an Kristallen. Berlin: Akademie-Verlag 1975

Kaiser, J.: Untersuchungen über das Auftreten von Geräuschen beim Zugversuch (An Investigation into the Occurrence of Noise in Tensile Tests). Dr.-Ing. Dissertation, TH München, 1950. English translation: UCRL-Trans-1082(L) Livermore June 1964

Kaminskii, A.A.; Belokoneva, E.L.; Mill, B.V.; Silvestrova, I.M.; Sarkisov, S.E.; Silvestrova, I.M.; Butashin, A.V.; Khodzhabagyan, G.G.: Phys. Status Solidi (a) 86 (1984) 345

Karcher, J.C.: A Piezoelctric Method for the Instantaneous Measurement of High Pressures. Jour. Franklin Inst., vol 194, pp 815-816, 1922

Kato, Y.; Nakamura, S.: On the Piezoelectric Accelerometer and Its Use in the Measurement of Elastic Waves Produced by Artificial Shocks. Sci. Reports Tohoku Univ., vol 19, ser. 1, pp 761-772, 1930; Proc. Imp. Acad. (Tokyo), vol 6, pp 272-274, 1930

Kawai, H.: The Piezoelectricity of Poly (Vinylidene Fluoride). Japan Journal of Applied Physics, vol 8, pp 975-976, 1969

Kenner, H.: Experimental Mechanics 15 (1975) 102

Keys, D.A.: Piezoelectric Method of Measuring Explosion Pressures. Phil. Mag. vol 42, pp 473-488, 1921

Kinigadner, A.; Kittinger, E.; Seil, K.; Tichý, J.: Die elektromechansichen Nichtlinearitäten von Quarzkristallen. Jahrestagung Oesterr. Phys. Ges. Loeben 1977

Kirchheim, A.; Sidler, A.; Stirnimann, J.: Zerspankraftmessung bei der Feinstbearbeitung, neues 3-Komponenten Dynamometer (Measurement of Cutting Forces in High-Precision Machining). Zeitschrift für wirtschaftlichen Fabrikbetrieb 6/96. (available as reprint 20.176e from Kistler)

Kishinouye, F.: Frequency Distribution of the Ito Earthquake Swarm of 1930. Bulletin of the Earthquake Research Institute, Tokyo Imperial Univ., 15 (part 2), pp 785-826 and plate LVL, Figs. 5 and 15

Kishinouye, F.: An Experiment on the Progress of Fracture (A Preliminary Report): Jishin, 6, 1934, pp 25-31 (in Japanese), Tokyo

Kishinouye, F.: An Experiment on the Progress of Fracture (A Preliminary Report). J. of Acoustic Emission, 9 (3), July/September 1990 (English translation of Kishinouye, 1934 by Ono Kanji, UCLA, 1989)

Kistler, W.P.: Messverstärker zur Messung von elektrischer Ladung. Schweiz. Patentsch. 267 431, Bern 1950

Klassen-Neklyudova, M.V.: Mechanical Twinning of Crystals (translated from Russian). New York: Consultants Bureau 1964

Kluge, J.; Linckh, H.E.: The Measurment of Compression- and Acceleration-forces by Piezoelectric Methods. Z. Verl. deut. Ing. vol 73, pp 1311-1314, 1929; see also vol 74, p 887, 1930; vol 75, p 115, 1931; also Z. tech. Mechanik Thermodynamik, vol 2, pp 153-164, 1931;

Arch. tech. Messen, vol 2, Lieferung 15, pp V132-133, 1932; Elektrotech. Z., vol 54, pp 158-159, 1933; Forsch. Gebiete Ingenw., vol 2, pp 153-164, 1931 (abst. in Electronics, vol 3, p 115, 1931 and vol 4, pp 177-182, 1933

Kochurikhin, V.V.; Kumatoriya, M.; Shimamura, K.; Takagi, H.; Fukuda, T.: J. Crystal Growth 181 (1997) 452

Kolodieva, S.V.; Firsova, M.M.: Soviet Physics-Crystallography 13 (1969) 540

Krempl, P.W.; Krispel, F.; Wallnöfer, W.: Industrial Development and Prospects of $GaPO_4$. Ann. chim. Sci. Mat, 1997, 22, pp 623-626

Krempl, P.W.; Schleinzer, G.; Wallnöfer, W.: Gallium phosphate, $GaPO_4$: a New Piezoelectric Crystal Material for High-Temperature Sensorics. Elsevier Sequoia: Sensors and Actuators A 61 (1997) 361-363

Krispel, F.; Krempl, P.W.; Knoll, P.; Wallnöfer, W.: OH Impurities in Gallium Orthophosphate. Proc. 11th Eur. Frequency and Time Forum, Neuchâtel: 4-6 March 1997, pp 66-70

Krispel, F.; Schleinzer, G.; Krempl, P.W.; Wallnöfer, W.: Measurement of the Piezoelectric and Electrooptic Constants of $GaPO_4$ with a Michelson Interferometer. Ferroelectrics, 1997, Vol. 202, pp 307-311

Kuratle, R.: Motorenmesstechnik. Würzburg: Vogel Buchverlag 1995

Landolt-Börnstein, Neue Serie (Hrsg. K. H. Hellwege), Bd. III/11. Elastische, piezoelektrische, pyroelektrische, piezooptische, elektrooptische Konstanten und nichtlineare dielektrische Suszeptibilitäten von Kristallen. Neubearbeitung und Erweiterung der Bände III/1 und III/2. Berlin, Heidelberg, New York: Springer 1979

Lang, S.B.: Bibliography on Piezoelectricity and Pyroelectricity of Polymers. Ferroelectrics, New York, Vol. 103 (1990), pp 219-337

Lang, S.B.: Guide to the Literature of Piezoelectricity and Pyroelectricity. Ferroelectrics, Amsterdam, Vol. 210, nos. 1/4 (1998), pp 83-336

Langeivn, P.: Sounding by Means of Sound Waves. Bur. hydrogr. internat. Spec. Pub. 3, Monaco, 1924

Langevin, P.; Chilowsky, C.: Echo Sounding. Nature, vol 115,pp 689-690, 1925; also Bur. hydrogr. internat. Spec. Pub. 14, Monaco, August, 1926

Liebertz, J.: Angew. Chemie 85 (1973) 326

Liebig, V.; Pridöhl, E.; Lohse-Koch, S.; Hoppe, N.; Cavalloni, C.; Kirchheim, A.: Absolute Calibration of Acoustic Emission (AE) Sensors with Laser Induced Surface Waves. Proceedings European Conference on Acoustic Emission Testing. Vienna, 6-8 May 1998, pp 240-248

Lippmann, M.G.: Ann. Chimie et Phys. 24 (1881) 145

Mahr, W.; Gautschi, G.H.: Piezotron Transducers – How they Compare with Piezoelectric Transducers Using Charge Amplifiers. London: Transducer' 82 Conf. 1982 (available as reprint 20.106e from Kistler)

Mandel, B.; Gautschi, G.: Sensors as Tools in Engine and Vehicle Engineering. ATZ/MTZ Sonderausgabe 1997/98. (available as reprint 20.186e from Kistler)

Mansfeld, G.D.; Boy, J.J.: JETP Lett. 66 (1997) 362

Marschall, K.; Gautschi, G.H.: In-Process Monitoring with Piezoelectric Sensors. J. of Materials Processing Technology, 44 (1991), pp 345-352

Martini, K.H.: Motortechn. Z. 25 (1964) 282

Martini, K.H.: Ölhydraulik und Pneumatik 15 (1971) 16

Martini, K.H.: Messen und Prüfen 7 (1971) 5 und 7 (1971) 50

Martini, K.H.: New Range of High-Temperature Quartz Pressure Transducers. London: Transducer '77 Conference 1977

Martini, K.H.: Multicomponent Dynamometers Using Quartz Crystals as Sensing Elements. ISA Trans, 22(1). ISA 1983 (available as reprint 20.183e from Kistler)

Martini, K.H.: Multicomponent Dynamometry on the Rolling Wheel. VDI Report No. 553, 1983. (available as reprint 20.122e from Kistler)

Mason, W.; McSkimin, H.J.; Shockley, W.: Ultrasonic Observation of Twinning in Tin. Physical Review, Vol. 73, No. 10 (May 1948), pp 1213-1214

Mason, W.P.: Piezoelectric Crystals and their Application to Ultrasonics. Toronto, New York, London: Van Nostrand 1950

Matsuda,Y.; Nagai, S.: Surface Wave Calibration of Acoustic Emission Sensors with Laser-Generated Ultrasound. J. Acoust. Soc. Jpn. (E), Vol. 15 (1994), 255

Mayer, G.: Recherches expérimentales sur une transformation du quartz. Diss. Univ. Paris 1959

McConnell, K.G.: Vibration Testing: Theory and Practice. New York: John Wiley & Sons 1995

McLaren, A.C.; Phakey, P.P.: Phys. Stat. Solid. 31 (1969) 723

McSkimin, H.J.; Andreatch, P.; Thurston, R.N.: J. Appl. Phys. 36 (1965) 1624

Mill, B.V.; Belokoneva, E.L.; Fukuda, T.: Zh. Neorg. Khim. 43 (1998) 1125 and 1270. (Translation in Russian J. Inorg. Chem. 43 (1998) 1032 and 1168)

Mill, B.V.; Fukuda, T.: Zh. Neorg. Khim. 43 (1998) 538. (Translation in Russian J. Inorg. Chem. 43 (1998) 470)

Mill, B.V.; Pisarevsky, Yu.V.; Belokoneva, E.L.: Proc. IEEE Int. Frequency Control Symposium, 1999, p 829

Mill, B.V.; Pisarevsky, Yu.V.: Langasite-Type Materials: from Discovery to Present State. Proc. 2000 IEEE/EIA International Frequency Control Symposium, 7–9 June 2000, Kansas City, Missouri, USA, pp 133–144

Millard, D.: Twinning in Single Crystals of Cadmium. PhD. thesis. Bristol, England: University of Bristol 1950

Miller, R.K.; McIntire, P.: Nondestructive Testing Handbook (2nd ed), vol 5: Acoustic Emission Testing. American Society for Nondestructive Testing 1987

Mitsui, T. et al: Ferro- und Antiferroelektrische Substanzen. Landolt-Börnstein, Neue Serie (Hrsg. K. H. Hellwege), Bd. III/9. Berlin, Heidelberg, New York: Springer 1975

Mitsui, T.; Marutake, M.; Sawaguchi, E.; et al: Ferro- und Antiferroelektrische Substanzen. Landolt-Börnstein, Neue Serie (Hrsg. K. H. Hellwege), Bd. III/3. Berlin, Heidelberg, New York: Springer 1969

Naumov, V.S.; Kalashnikova, I.I.; Aleshko-Ozhevskii, O.P.: Proc. IEEE Int. Frequency Control Symposium, 1996, p 122

Neubert, H.K.P.: Instrument Transducers, 2nd ed Oxford: Clarendon Press 1975

Newnham, R.E.: Am. Mineralogist 69 1974) 906

Newnham, R.E.: Structure-Property Relations. Berlin, Heidelberg, New York: Springer 1975

Newnham, R.E.; Cross, L.E.: Mat. Res. Bull. 9 (1974) 927; 9 (1974) 1021

Nitsche, W.; Mirow, P.; Szodruch, A.: Piezo-Electric Foils as a Mean of Sensing Unsteady Forces on Flow-Around Bodies. Sixth Symposium on Turbulent Shear Flows, September 1987, Toulouse France

Nitsche, W.; Mirow, P.; Deulitis, A.: Einsatzmöglichkeiten piezoelektrischer Folien zur Messung von Oberflächenkräften. Proc. Sensor 88, Nürnberg, May 1988

N.N.: Piezoelectricity (Selected Engineering Reports). London: Her Majesty's State Office 1957

N.N.: Materials for High Temperature Acoustic and Vibration Sensors: a Review. Applied Acoustics, 41 (1994), 299

N.N.: Endevco Dynamic Test Handbook (a Compendium of Formulae, Tables, Conversion Factors, and Other Technical Data Frequently Used in Dynamic Testing). San Juan Capistrano CA: Endevco 1997

N.N.: High Quality Components & Materials for the Electronic Industry (Product catalog). Kvistgård (DK-3490): Ferroperm A/S, 2001

Noltingk, B.E. (ed): Instrumentation Reference Book. Oxford: Butterworth-Heinemann 1996

Norton, H.N.: Handbook of Transducers for Electronic Measuring Systems. Englewoof Cliffs: Prentice Hall 1969

Norton, H.N.: Sensor and Analyzer Handbook. Englewood Cliffs: Prentice Hall 1982

Okochi, M., Hashimoto, S.; Matsui, S.: High Speed Internal Combustion Engine and Piezoelectric Pressure Indicator. Bull. Inst. Phys. Chem. Research (Tokyo), vol 4, pp 85-97, 1925

Okochi, M.; Okoshi, M.: New method for Measuring the Cutting Force of Tools and Some Experimental Results. Sci. Papers Inst. Phys. Chem. Research (Tokyo), vol 5, pp 261-301, 1927; see also vol 12, pp 167-192, 1930

Okochi, M.; Okoshi, M.: Deflection of a Coninuous Beam and Force Exerted on the Support by a Moving Load. Bull. Inst. Phys. Chem. Research (Tokyo), vol 7, pp 659-667, 1928

Okochi, M.; Miyamoto, T.: Balancing Machine Utilizing Piezoelectricity. Bull. Inst. Phys. Chem. (Tokyo), vol 7, pp 383-391, 1928; see also vol 9, pp 6-10, 1930

Palmier, D.; Goiffon, A.; Gohier, R.; Fischer, M.; A. Zarembovich, A.; Philippot, E.: Elastic Behavior Under Hydrostatic Pressure and Acoustic-Mode Vibrational Anharmonicity of Single-Crystal Gallium Phosphate ($GaPO_4$). Ann. Chim. Sci. Mat. 22, 627-632 (1997)

Patel, N.D.; Nicholson, P.S.: High-Frequency, High-Temperature Ultrasonic Transducers. NDT International, 23(5), 1990, p 262

Peggs, G.N. (ed): High Pressure Measurement Techniques. London and New York: Applied Science Publishers Ltd. 1983

Pennington, D.: Piezoelectric Accelerometer Manual (company publication). Pasadena: Endevco 1965

Petrzilka, V.; Slavik, J.B.; Solc, I.; Taraba, O.; Tichý, J.; Zelenka, J.: Piezoelectricity and its Technical Applications (in Czech). Prague: Academia 1960

Pflier, P.M.; Jahn, G., Jentsch, G.: Elektrische Messgeräte und Messverfahren, 4. Auflage. Berlin, Heidelberg, New York: Springer 1978

Philippot, E.; Ibanez, A.; Goiffon, A.; Cochez, M.; Zarka, A.; Capelle, B.; Schwartzel, J.; Détaint, J.: A Quartz-Like Material: Gallium Phosphate ($GaPO_4$); Crystal Growth and Characterization. J. Crystal Growth 130 (1993) 195-20

Philippot, E.; Pisarevsky, Yu.V.; Capelle, B.; Détaint, J.: Present State of the Development of the Piezoelectric Materials. Proc. 15th European Frequency and Time Forum, EFTF 2001, 6–8 March, 2001, Neuchâtel, Switzerland, pp 33–37. (ed: by FSRM, Swiss Foundation for Research in Microtechnology, Neuchâtel)

Piotti, M.R.; Sirtes G.C.: Drei-dimensionale Kraftmessungen bei implantatverankerten Hybridprothesen – eine vergleichende In-vivo-Studie mit verschiedenen Verankerungs-elementen im zahnlosen Unterkiefer (3-dimensional Force Measurements in Implant-Anchored Hybrid Prostheses – a Comparative In-Vivo Study with Different Anchoring Elements in a Toothless Jaw). PhD Thesis Univ. Bern, 1996

Pisarevsky, Yu.V.; Senushencov, P.A.; Popov, P.A.; Mill, B.V.: Proc. IEEE Int. Frequency Control Symposium, 1995, p 653

Pisarevsky, Yu.V.; Senyushenkov, P.A.; Mill, B.V.; Moiseeva, N.A.: Proc. IEEE Int. Frequency Control Symposium, 1998, p 742

Plankey, R.; Wilson, J.: Shock and Vibration Measurement Technology. Endevco Corp. USA: 2001

Polla, D.L.; Francis L.F.: Ferroelectric Thin Films in Microelectromechanical Systems Applications. MRS Bulletin, July 1996, p 59

Profos, P.; Pfeifer, T. (Hrsg.): Handbuch der industriellen Messtechnik. München: Oldenbourg 1994

Profos, P.; Pfeifer, T. (Hrsg.): Grundlagen der Messtechnik. München: Oldenbourg 1997

Raaz, F.: Röntgenkristallographie. Berlin: de Gruyter & Co. 1975

Raman, C.V.; Nedungadi, T.M.K.: Nature 145 (1940) 147

Randeraat, J. van; Setterington, R.E. (Publ.): Piezoelectric Ceramics, 2nd ed London: Mullard 1974

Reiter, C.; Thanner, H.; Wallnöfer, W.; Krempl, P.W.: Properties of $GaPO_4$ Thickness Shear Resonators. Ann. Chim. Sci. Mat. 22, 633-6 (1997)

Reiter, C.; Thanner, H.; Wallnöfer, W.; Worsch, P.M.; Krempl, P.W.: Elastic Constants and Temperature-Compensated Orientations of $GaPO_4$. Proc 15. Eur. Frequency and Time Forum, 2001)

Reiter, C.; Krempl, P.W.; Thanner, H.; Wallnöfer, W.; Worsch, P.M.: Material Properties of $GaPO_4$ and their Relevance for Applications. Proc. 3rd Eur. Workshop on Piezoelectric Materials, Montpellier: 5-6 October 2000

Riecke, E.; Voigt, W.: Wied. Ann. 45 (1892) 523
Roberts, H.C.: Mechanical Measurements by Electrical Methods. Pittsburgh: Instrument Publ. 1946
Rohrbach, Chr.: Handbuch für elektrisches Messen mechanischer Grössen. Düsseldorf: VDI-Verlag 1967
Rosen, C.Z.; Hiremath, B.V.; Newnham, R.E. (eds): Piezoelectricity. New York: American Institute of Physics 1992
Sayer, M.; Judd, B.A.; EI-Assal, K.; Prasad, E.: Poling of Piezoelectric Ceramics. J. Canad. Ceramic Soc. vol 50, 1981
Schweppe, H.; Quadflieg, P.: IEEE Transactions on Sonics and Ultrasonics SU-21 (1974) 56
Seil, K.: Elektroelastische Eigenschaften von ADP und Turmalin. Diss. Naturwiss. Fak. Univ. Innsbruck 1978
Selbstein, U.: Applications industrielles des mesures électroniques. Paris: Dunond 1950
Sheludew, I.S.: Elektrische Kristalle. Berlin: Akademie-Verlag 1975
Shorrocks, N.M.; Whatmore, R.W.; Ainger F.W. and Young I.M.: Proc. IEEE Ultrasonics Symposium, 1981, p 337
Shubnikov, A.V.: Piezoelectric Textures. Moskwa: Izd. Akad. Nauk SSSR 1946
Shubnikov, A.V.; Zheludev, I.S.; Konstantinova, V.P.; Silvestrova, I.M.: Issledovanie piezoelektriceskih texstur. Moskwa: Izd. Akad. Nauk SSSR 1956
Silvestrova, I.M.; Pisarevsky, Yu.V.; Mill, B.V.; Belokopitov, A.V.; Moiseeva, N.A.: Proc. 14th European Frequency and Time Forum, Torino, Italy, 2000
Silvestrova, I.M.,; Pisarevsky, Yu.V.; Kaminskii, A.A.; Mill, B.V.: Fiz. Tverd. Tela (Leningrad) 29 (1987) 1520. (Translation in Sov. Phys. Solid State 29 (1987) 870)
Spur, G.; Al-Badrawy, S.J.; Stirnimann, J.: Zerspankraftmessung bei der fünfachsigen Fräsbearbeitung (Measuring Cutting Force in Five-Axis Milling). Zeitschrift für wirtschaftliche Fertigung und Automatisierung 9/93. München: Carl Hanser Verlag 1993. (available as reprint 20.162e from Kistler)
Ssakharov, S.A.; Larionov, I.M.; Medvedev, A.V.: Proc. IEEE Int. Frequency Control Symposium, 1992, p 713
Stirnimann, J.; Kirchheim, A.: New Cutting Force Dynamometers for High-Precision Machining. Industrial Tooling '97, 9/10 September 1997 (available as reprint 20.185e from Kistler)
Sydenham, P.H.: Introduction to Measurement Science and Engineering. P.H. Sydenham, N.H. Hancock and R. Thorn. Chichester: Wiley 1989
Sydenham, P.H.; Hancock, N.H.; Thorn, R.: Handbook of Measurement Science. Chichester: John Wiley and Sons 1992
Taylor, G.W. (ed): Piezoelectricity. New York: Gordon & Breach 1985, pp 231-234
Thanner, H.; Krempl, P.W.; Krispel, F.; Reiter, C.; Selic, R.: High Temperature Microbalance Based on $GaPO_4$. Proc. 3rd Eur. Workshop on Piezoelectric Materials, Montpellier: October 5-6, 2000
Thomas, L.A.; Rycroft, J.L.: Nature 157 (1946) 406
Thomas, L.A.; Wooster, W.A.: Proc. Royal Soc. (London) A 208 (1951) 43
Thomson J.J.: Piezoelectricity and its Application (Quartz and Tourmaline for Explosion Pressures). Engineering, vol 107, pp 543-544, 1919
Thurston, R.N.: J. Appl. Phys. 37 (1966) 1
Thurston, R.N.; Brugger, K.: Phys. Rev. 133 (1964) A 1604:
Thurston, R.N.; McSkimin, H.J.; Andreatch, O. Jr.: J. Appl. Phys. 37 (1966) 267
Tichý, J.; Gautschi, G.: Piezoelektrische Messtechnik. Berlin, Heidelberg, New York: Springer 1980
Tokarev, E.F.; Kobyakov, I.B.; Kuzmina, I.P.; Lobachev, A.N.; Pado, G.S.: Fiz. Tverd. Tela 17 (1975) 980. (Translation in Sov. Phys. Solid State 17 (1975) 629)
Tschappat, W.H.: New Instruments for Physical Measurements (pressure gauge for interior ballistics). Mech. Eng., vol 45, pp 673-678, 1923; see also vol 48, pp 821-825, 1926
Tschapek, M.; Santamaria, K.; Natale, I.: Electrochimica Acta 14 (1969) 889

Tsubouchi, K.; Sugai, K.; Mikoshiba, M.: AlN Material Constants Evaluation and SAW Properties on AlN/Al2O3 and AlN/Si. 1981 Ultrasonics Symposium, 1981, p 375
Uchino, K.: Ferroelectric Devices. New York, Marcel Dekker 2000
Voigt, W.: Lehrbuch der Kristallphysik. Leipzig, Berlin: Teubner 1910
Voisard, C.: Conductivity, Dielectric and Piezoelectric Properties of $SrBi_4Ti_4O_{15}$. Thesis Ecole Polytechnique Fédérale de Lausanne (Switzerland): June 2000
Wallnöfer, W.; Krempl, P.W.; Asenbaum, A.: Determination of Elastic and Photoelastic Constants of Quartz-Type $GaPO_4$ by Brillouin Scattering. Physical Review B, Vol. 49, No. 15, April 1994, pp 10075-10080
Wallnöfer, W.; Krempl, P.W.; Krispel, F.: Growth Zones and their Boundaries in $GaPO_4$. Proc. 3rd Eur. Workshop on Piezoelectric Materials, Montpellier: 5-6 October 2000
Watanabe, S.: A New Design of Cathode Ray Oscillograph and Its Applications to Piezoelectric Measurements. Sci. Papers Inst. Phys. Chem. Research (Tokyo), vol 5 , pp 261-301, 1927; see also vol 12, pp 82-98, 1929; and Proc. World Eng. Congr., Tokyo, vol 5, Paper No. 632
Webster, John G. (Ed.): The Measurement, Instrumentation and Sensors Handbook (The Elec. Engineering Handbook Series). 1999. Boca Raton, FL: CRC Press 1999
White, D.L.: J. Acoust. Soc. Am. 11 (1959) 311
Wood, H.O.: On a Piezoelectric Accelerograph. Bull. Seismol. Soc. Am., vol 11, pp 15-57, 1921
Wooster, W.A.; Wooster, N.: Nature 157 (1946) 405
Wooster, W.A.; Wooster, N.; Ryecroft, I.L.; Thomas, L.A.: J. Elec. Engs. 94 (Pt. III) (1947) 926
Worsch, P.M.; Koppelhuber-Bitschnau, B.; Mautner, F.A.; Wallnöfer, W.: Temperature Dependence of the Alpha-Galliumphosphate Structure. Materials Science Forum, Vol. 278-281 (1998) 600
Worsch, P.M.; Thanner, H.; Krempl, P.W.; Krispel, F.; Reiter, C.; Wallnöfer, W.: The High-Temperature Piezoelectric Material $GaPO_4$ and its Application for Microbalance Sensors. Proc. SENSOR 01 Conference, Nürnberg, Germany, 8-10 May 2001, paper A6.2
Wun-Fogle, M. et al (eds): Materials for Smart Systems. III Symposium 30 November - 2 December 1999, Boston. Warrendale PA : Materials Research Society 2000
Xu, Y.: Ferroelectric Materials and their Applications. North Holland 1991
Yamaguchi, K.: An Accelerometer Utilizing Piezoelectricity. Bull. Inst. Phys. Chem. Research (Tokyo), vol 8, pp 157-163, 1929
Zelenka, J.: Czech J. Phys. B 28 (1978) 165
Zelenka, J.; Lee, O.C.: IEEE Trans. on Sonics and Ultrasonics SU-18 (1971) 79
Zeller, W.: Experimental and Theoretical Investigation of Vibration Meters for Traffic Vibrations. Z. Bauwesen, vol 80, pp 171-184, 1930

List of Manufacturers

Manufacturers of piezoelctric sensors (as of May 2001, S.E. & O.)

AE Sensors BV	Phone	+ 31 - 78 - 621 31 52
Jan Valsterweg 92, PO Box 9084	Fax	+ 31 - 78 - 621 31 46
NL-3301 AB Dordrecht, The Netherlands	Internet	www.aesensors.nl
AVL List GmbH	Phone	+ 43 - 316 787-0
Hans-List-Platz 1	Fax	+ 43 - 316 787-400
AT-8020 Graz, Austria	Internet	www.avl.com
Baumer sensopress AG	Phone	+ 41 - 52 - 728 13 93
Hummelstrasse 17	Fax	+ 41 - 52 - 728 13 95
CH-8501 Frauenfeld, Switzerland	Internet	www. baumersensopress.com
Beijing Vibro-Measuring Instrument Factory	Phone	+ 86 - 10 - 63 52 18 82
8 Baiguang Lu, Xuanwu District	Fax	+ 86 - 10 - 63 52 36 54
Beijing 100053, PRC		
Bently Nevada Corporation	Phone	+ 1 - 775 - 782 36 11
1631 Bently Parkway South	Fax	+ 1 - 775 - 782 92 53
Minden, NA 89423, USA	Internet	www.bently.com
Beran Instruments Ltd.	Phone	+ 44 - 18 05 62 43 04
Hatchmoor Industrial Estate	Fax	+ 44 - 18 05 62 40 93
Torrington, Devon EX38 7HP, UK	Internet	www.beran.co.uk
D.J. Birchall Ltd,	Phone	+ 44 - 1638 - 71 42 24
Finchley Avenue	Fax	+ 44 - 1638 - 71 75 31
Mildenhall, IP28 7BG, UK	Internet	www.djbirchall.com
Brüel & Kjær (HQ)	Phone	+ 45 45 80 05 00
Skodsborgvej 307	Fax	+ 45 45 80 14 05
DK-2850 Nærum, Denmark	Internet	www.bksv.com
Columbia Research Laboratories, Inc.	Phone	+ 1 - 800 - 813 84 71
1925 Mac Dade Boulevard	Fax	+ 1 - 610 - 872 38 82
Woodlyn, PA 19094, USA	Internet	www.columbiaresearchlab.com

Dunegan Engineering Consultants, Inc.	Phone	+ 1 - 949 - 661 81 05
PO Box 1749	Fax	+ 1 - 949 - 661 37 23
San Juan Capistrano, CA 92693, USA	Internet	www.deci.com
Dunegan-Trodyne	Phone	+ 1 - 609 - 716 41 90
PO Box 206	Fax	+ 1 - 609 - 716 41 91
West Windsor, NJ 08550-5303	Internet	www.acousticemission.com
Dynasen, Inc.	Phone	+ 1 - 805 - 964 44 10
20 Arnold Place	Fax	+ 1 - 805 - 967 28 24
Goleta, CA 93117, USA	Internet	www.dynasen.com
Dytran Instruments, Inc.	Phone	+ 1 - 818 - 700 78 18
21592 Marilla Street	Fax	+ 1 - 818 - 700 78 80
Chatsworth, CA 91311, USA	Internet	www.dytran.com
Endevco Corporation	Phone	+ 1 - 949 - 493 81 81
30700 Rancho Viejo Road	Fax	+ 1 - 949 - 661 72 31
San Juan Capistrano, CA 92675, USA	Internet	www.endevco.com
IMCO International Ltd.	Phone	+ 1 - 815 - 469 25 00
245 East Laraway Road	Fax	+ 1 - 815 - 469 60 09
Frankfort, IL 60423, USA	Internet	www.loadmonitors.com
Kistler Instrumente AG Winterthur	Phone	+ 41 - 52 - 224 11 11
Eulachstrasse 22, PO Box 304	Fax	+ 41 - 52 - 224 14 14
CH-8408 Winterthur, Switzerland	Internet	www.kistler.com
Ktech Corp.	Phone	+ 1 - 505 - 998 58 30
2201 Buena Vista SE, Suite 400	Fax	+ 1 - 505 - 998 58 50
Albuquerque, NM 87106-4265	Internet	www.ktech.com
Lattron Co., Ltd.	Phone	+ 82 - 42 - 935 84 32
1688-24 Shinil-dong, Daedeok-gu	Fax	+ 82 - 42 - 935 20 34
Taejeon 306-230, Rep. of Korea	Internet	www.lattron.co.kr
Measurement Specialties, Inc.	Phone	+ 1 - 610 - 650 15 00
950 Forge Avenue - Bldg B	Fax	+ 1 - 610 - 650 15 09
Norristown, PA 19403, USA	Internet	www.msiusa.com/contact.htm
Nordisk Transducer Teknik	Phone	+ 45 - 98 58 14 44
Als Odde	Fax	+ 45 - 98 58 18 66
DK-9560 Hadsund, Denmark	Internet	www.transducertechnic.com
Ono Sokki Co., Ltd.	Phone	+ 81 - 45 - 935 39 76
1-16-1 Hakusan, Midori-ku	Fax	+ 81 - 45 - 930 19 06
Yokohama 226-8507, Japan	Internet	www.onosokki.co.jp
PCB Piezotronics, Inc.	Phone	+ 1 - 716 - 684 00 01
3425 Walden Avenue	Fax	+ 1 - 716 - 684 09 87
Depew, NY 14043-2495, USA	Internet	www.pcb.com

Physical Acoustics Corporation 195 Clarksville Road Princeton Junction, NJ 08550-5303, USA	Phone Fax Internet	+ 1 - 609 - 716 40 00 + 1 - 609 - 716 07 06 www.pacndt.com
RDP Electronics, Ltd. Grove Street Heath Town, Wolverhampton, WV10 0PY, UK	Phone Fax Internet	+ 44 - 1902 - 45 75 12 + 44 - 1902 - 45 20 00 www.rdpelectronics.com
Sensotech, Inc. 2080 Arlingate Lane Columbus, OH 43228, USA	Phone Fax Internet	+ 1 - 614 - 850 50 00 + 1 - 614 - 850-11 11 www.sensotec.com
TEAC Corp. 3-7-3 Naka-cho, Musashino-shi Tokyo 180-8550, Japan	Phone Fax Internet	+ 81 - 422 - 52 50 00 + 81 - 422 - 55 89 59 www.teac.co.jp/ipd/dr/transducere
Toray Techno Co., Ltd. 1-1-1 Sonoyama Otsu-shi, Shiga-ken, 520-8558, Japan	Phone Fax Internet	+ 81 - 77 - 537 51 88 + 81 - 77 - 534 87 24 www.toraytechno.ab.psiweb.com
Vallen-Systeme GmbH Schäftlarner Weg 26 DE-82057 Icking, Germany	Phone Fax Internet	+ 49 - 8178 - 96 74-400 + 49 - 8178 - 96 74-444 www.vallen.de
Vibrametrics, Inc. 1014 Sherman Avenue Hamden, CT 06514, USA	Phone Fax Internet	+ 1 - 203 - 288 61 58 + 1 - 203 - 288 49 37 www.vibrametrics.com
Vibro-Meter SA Route de Moncor 4, PO Box 1071 CH-1701 Fribourg, Switzerland	Phone Fax Internet	+ 41 - 26 - 407 11 11 + 41 - 26 - 407 13 01 www.vibro-meter.com
Wilcoxon Research, Inc. 21 Firstfield Road Gaithersburg, ML 20878, USA	Phone Fax Internet	+ 1 - 301 - 330 88 11 + 1 - 301 - 330 88 73 www.wilcoxon.com
Xi'an Zhongxing Test Control Co., Ltd. F8, TowerB, New Energy Bldg., Nanyou Road Shenzhen, Guangdong 518054, PRC	Phone Fax Internet	+ 86 - 755 - 664 11 03 + 86 - 755 - 664 59 76 www.66daji.com
Yangzhou No. 2 Radio Factory 1 Yuqi Street Yangzhou 225002, Jiangsu Province, PRC	Phone Fax Internet	+ 86 - 514 - 734 87 17 + 86 - 514 - 734 86 70 www.china-yec.com

Index

GPSR Compliance
The European Union's (EU) General Product Safety Regulation (GPSR) is a set of rules that requires consumer products to be safe and our obligations to ensure this.

If you have any concerns about our products, you can contact us on

ProductSafety@springernature.com

In case Publisher is established outside the EU, the EU authorized representative is:

Springer Nature Customer Service Center GmbH
Europaplatz 3
69115 Heidelberg, Germany

www.ingramcontent.com/pod-product-compliance
Ingram Content Group UK Ltd.
Pitfield, Milton Keynes, MK11 3LW, UK
UKHW021832190726
13853UKWH00003B/1280

* 9 7 8 3 6 6 2 0 4 7 3 3 0 *